Statistical Methods
for Chemists

WILEY PUBLICATIONS IN STATISTICS

Walter A. Shewhart, Editor

Mathematical Statistics

    DWYER—Linear Computations.

    FISHER—Contributions to Mathematical Statistics.

    WALD—Statistical Decision Functions.

    FELLER—An Introduction to Probability Theory and Its Applications, Volume One.

    WALD—Sequential Analysis.

    HOEL—Introduction to Mathematical Statistics.

Applied Statistics

    YOUDEN—Statistical Methods for Chemists.

    KEMPTHORNE—Design and Analysis of Experiment (*in press*).

    HALD—Statistics (*in press*).

    MUDGETT—Index Numbers.

    TIPPETT—Technological Applications of Statistics.

    DEMING—Some Theory of Sampling.

    COCHRAN and COX—Experimental Designs.

    RICE—Control Charts.

    DODGE and ROMIG—Sampling Inspection Tables.

Related Books of Interest to Statisticians

    HAUSER and LEONARD—Government Statistics for Business Use.

Statistical Methods for Chemists

W. J. YOUDEN
National Bureau of Standards
Washington, D. C.

New York · John Wiley & Sons, Inc.
London · Chapman & Hall, Limited

COPYRIGHT, 1951
BY
JOHN WILEY & SONS, INC.

All Rights Reserved

This book or any part thereof must not be reproduced in any form without the written permission of the publisher.

COPYRIGHT, CANADA, 1951, INTERNATIONAL COPYRIGHT, 1951
JOHN WILEY & SONS, INC., PROPRIETORS

All Foreign Rights Reserved

Reproduction in whole or in part forbidden.

PRINTED IN THE UNITED STATES OF AMERICA

PREFACE

THIS BOOK IS WRITTEN FOR THOSE WHO MAKE MEASUREMENTS AND interpret experiments. The book is characterized by an absence of statistical theory and proofs. I am convinced, from observation, that young men who have been given a fine mathematical grounding in statistics often have an inadequate grasp of the ways in which measurements behave. Observation has equally shown that most experimental workers have a very real understanding of the meaning of their measurements. They are, unfortunately, inarticulate. The purpose of this book is to make available to the scientist the modern statistical system of units for expressing scientific conclusions.

Many statistical techniques require no demonstration for acceptance by the experimental scientist. The scientist instinctively sees that the exact formulations of the statistician are in accord with his experience. His confidence in statistical verdicts rests more upon this accord than upon mathematical derivations.

The claim is frequently made that there will be many misapplications of statistical techniques if the user lacks a solid background in the derivation of statistical theorems. Of course it is also possible for the mathematical statistician to go astray if his knowledge of science is inadequate. Misapplications there will be. Many biologists use chemical techniques, and sometimes their work is open to criticism. I believe that much less would have been achieved in biological work if the use of chemical procedures had been restricted to biologists who also possessed a degree in chemistry.

Examination of the chapters shows (i) that a good deal of the usual introductory material is missing; (ii) that some elementary topics have been given detailed consideration and others discussed only sufficiently to attract attention to them; and (iii) that several topics usually included in a statistics textbook are omitted.

The introductory material has been held to the minimum, because I believe that laboratory men know a good deal about data that is not known to students taking a school course in statistics. The omissions reflect my conviction that many of the ordinary situations confronting the scientist require knowledge of only a few statistical techniques.

In the later chapters I have tried to reveal the essential characteristics of some useful experimental arrangements. When senior workers, who are responsible for the direction of investigations, realize the possible gains from these arrangements they will encourage younger men to go more fully into the subject of experimental design. The selection, indeed the invention, of new designs properly belongs in the hands of those who know best the circumstances in which they are to be used.

W. J. YOUDEN

National Bureau of Standards
Washington, D. C.
July, 1951

ACKNOWLEDGMENTS

I AM PARTICULARLY INDEBTED TO THE FOLLOWING MEMBERS OF THE staff of the National Bureau of Standards for making available unpublished data for a number of the examples:

D. N. Evans, Concreting Materials Section
F. W. Schwab, Inorganic Chemistry Section
B. F. Scribner, Spectrochemistry Section
H. H. Seliger, Radioactivity Section
R. E. Wilson, Temperature Measurements Section

L. Tanner, Chief Chemist, U. S. Customs, generously made available the analytical results on individual cores from baled wool.

Dr. D. B. DeLury, Ontario Research Foundation, assisted me in the section on degrees of freedom.

Grateful acknowledgment is made to R. A. Fisher and Frank Yates for permission to reprint parts of Tables III and V from their *Statistical Tables*.

The prompt appearance of the book is the result of the careful preparation of the copy for the printer by Miss Vivian Frye and Miss J. M. Jackson.

My knowledge of statistics stems directly from my statistical friends and their works. Any merit the book has must rest upon the selection and presentation of the material.

> When 'Omer smote his bloomin' lyre,
>   He'd 'eard men sing by land and sea;
> An' what he thought 'e might require,
>   'E went an' took—the same as me!
> —KIPLING

CONTENTS

CHAPTER
1. PRECISION AND ACCURACY ... 1
   Interpretation of data ... 1
   Small sets of data ... 3
   Independence of measurements ... 6
   Constant errors ... 6
   Rounding off data ... 7

2. THE MEASUREMENT OF PRECISION ... 8
   The average deviation ... 8
   The standard deviation ... 10
   Normal law of error ... 11
   Calculation of standard deviation ... 12
   Degrees of freedom ... 13
   Confidence limits ... 18
   Comparison of standard deviations ... 20

3. THE COMPARISON OF AVERAGES ... 24
   The t test ... 24
   The F test ... 29

4. THE RESOLUTION OF ERRORS ... 33
   Sampling and analytical errors ... 33
   Sampling in two stages ... 37

5. STATISTICS OF THE STRAIGHT LINE ... 40
   Detection of constant errors ... 40
   Standard deviation of slope ... 42
   Standard deviation of intercept ... 43
   Measuring the fit of the line ... 45
   Comparison of slopes ... 47

6. THE ANALYSIS OF VARIANCE ... 50
   Relation between t and F tests ... 50
   Use of variance to compare averages ... 53
   Precautions in using the analysis of variance ... 55

7. INTERACTION BETWEEN FACTORS ... 59
   The concept of interaction ... 59
   Interpretation of data ... 61

CONTENTS

CHAPTER
8. REQUIREMENTS FOR DATA — 72
   Planning the experimental program — 72
   Replication of measurements — 72
   Arrangement of experimental material — 74
   Random arrangements — 75

9. ARRANGEMENTS FOR IMPROVING PRECISION — 80
   Grouping measurements in blocks — 80
   Latin square arrangements — 90
   Incomplete block arrangements — 99

10. EXPERIMENTS WITH SEVERAL FACTORS — 106
    Factorial experiments — 106
    Confounding — 107
    Systematic selection of experimental trials — 112

LIST OF PUBLICATIONS REFERRED TO IN THE TEXT — 115

APPENDIX
   Table   I. Critical Values of t — 119
   Table II. Critical Values of F at 5 Per Cent Level — 120
   Table III. Critical Values of F at 1 Per Cent Level — 121
   Table IV. Table of Squares — 122

INDEX — 125

CHAPTER 1

Precision and Accuracy

Interpretation of Data

It is often thought that the statistician can add nothing to the experimenter's own interpretation of his measurements. Under certain circumstances the experimenter can interpret his data as well as or even better than the statistician. Suppose that the experimenter obtains a single measurement, for example 5.32, the result of a quantitative determination for copper. Without further information no statistician would hazard an estimate of how closely a second determination would agree with the first result. The experimenter, on the other hand, may have obtained the result by an analytical procedure which he has used many times. He has observed that his duplicates seldom differ by more than 0.06 when about 5 per cent of the element is present. He would consequently expect the second determination to fall in the range 5.26 to 5.38. He also knows that most of the time the second determination will fall within narrower limits, say 5.29 to 5.35. The experimenter possesses information not known to the statistician, and, in his own way, he makes use of this information.

The above example does not show that the experimenter has no need for statistics. Rather, the example shows the experimenter using statistical concepts, albeit in a somewhat primitive manner. If the experimenter made available to the statistician the further information concerning the agreement of duplicate measurements, the statistician would undertake to make some predictions about the next measurement. There are various ways in which this information about duplicate measurements may be expressed. The statement may be made that, on the average, only once in 20 times does the difference between a pair of satisfactory determinations exceed 0.06. A large number of pairs must be inspected to establish the limit which is exceeded by 5 per cent of the differences between the duplicates. It is better to report the average difference, 0.025, between duplicates. Twenty to thirty differences are sufficient to give a fairly satisfactory estimate of the average difference. If the statistician possesses information in either of these forms, he will be willing to go into considerable detail in his prediction about the second measurement. These predictions are often made in the following form:

> one in ten of the differences will exceed 0.05
> one in twenty of the differences will exceed 0.06
> one in a hundred of the differences will exceed 0.08

(assuming an average difference between duplicates of 0.025).

Leaving aside for the moment the question as to whether it is useful to be able to write down a prediction in such detail, few will find it possible to make such a tabulation without using a statistical table. The statements may be based upon the inspection of a large number of differences or upon calculation from the average difference. The average difference may be obtained from a relatively small number of pairs. The experienced investigator knows that small differences between duplicates occur more often than do large differences. The statistical operation to obtain the foregoing tabulation is something like the focusing operation with a camera lens whereby hazy outlines are converted into a sharp image. When the necessary information is given, statistical techniques produce clearer pictures of the event being predicted.

The ability to make sharper prediction is not the only advantage possessed by the worker who has acquired an understanding of a few elementary statistical ideas. There is another and more important benefit. This may be illustrated by recalling the two alternative ways of reporting information about the differences between duplicate determinations. If twenty sets of duplicates are available for examination, the next to largest difference may be reported together with the statement that only one of the twenty differences exceeded this value. The other alternative is to report the average of the twenty differences. It is a statistical problem to find out which is the more useful of these two pieces of information. It turns out that a better prediction can be made using the average difference.

There are other operations which might be performed on the twenty differences, such as taking the average of the squared differences. Given a definite and limited amount of experimental data, it is clear that it is desirable to choose a property of differences which will be most effective for making predictions.

The importance of making a prediction is sometimes overlooked, even in the simple case of predicting by how much a second determination may differ from the first result. When duplicates are run, disagreement between them may mean that something has gone wrong with one or the other of the determinations. If this is judged to be the case, a third analysis will be required to determine which one of the first two determinations is doubtful. The decision to make the third analysis rests solely on the fact that the difference between the duplicates exceeds an acceptable amount. This is simply saying that a limiting value for the difference has been chosen which, it is predicted, will occur or be exceeded only rarely, if both determinations are, in fact, good determinations. This is precisely a prediction, although very often the limit is not set in advance of looking at the results. In all fairness the prediction should be made first; otherwise there is a natural tendency to adjust the acceptable limit to the results in hand. The limit, when it is set in advance, is often set somewhat arbitrarily. If it is set too small, then a good many unnecessary third determinations will be made. If it is set too large, then numerous determinations which are in serious error may be retained in the work. All this indicates that it is worth while to make sharp predictions of the type tabulated above, and that it is helpful to be able to do this with a minimum of past experience. The achievement of these ends is a statistical task. The experimenter who says that he has no use for statistics does not mean that he ignores the problems just discussed. He only reveals that he is content to interpret his results intuitively and inefficiently without availing himself of the assistance of statistical techniques which have been developed for these problems.

It should be clearly understood that statistical techniques for the interpretation of data are usually devised to make use of the data as the sole source of information. The statistical formulas, as a rule, make no provision for utilizing accumulated experience from data other than the set under examination. Provision could be made to do this, but it is necessary to be quite sure that the prior experience is completely applicable to the data in hand.

Small Sets of Data

Small amounts of data can be used as a basis for interpretation and prediction. It is generally the case that small amounts of data lead to cautious interpretations that are of little practical value. If one pair of duplicate determinations is exhibited to a statistician and he is asked to predict the average difference between such sets of duplicates, he will be forced to report that the average difference may be anywhere from $\frac{1}{3}$ to 10 times the value of the one difference before him. Fortunately these brackets draw together rapidly as the number of pairs available for examination increases. Statistical processes mirror the feelings of an experimenter engaged in an entirely new field of work. Much latitude must be allowed in all estimates until the quantity of data accumulated is sufficient to reveal the situation. If this is so, how can statistics contribute anything? By means of statistical techniques the requisite latitude in making an estimate can be specified in terms of the quantity of data available. Very often it is not necessary to work to close limits. Most problems have their own requirements in this respect. Statistical formulas do supply a simple means of evaluating the amount of work required to be done. Statistical verdicts may be regarded as mathematically objective judgments based upon the numerical relationships among measurements. Statistical formulas are internationally accepted and, like a table of international atomic weights, are useful in avoiding unnecessary disputes.

The limitations of small amounts of data are shown very well in connection with the rejection of observations. Every now and then a measurement is very wide of the mark. This situation usually arises through some oversight of the worker, such as the gross misreading of a scale or weights, or the transposition of figures in recording the result. All workers are interested in detecting the presence of such mishaps in their work. Indeed the prevailing custom of performing duplicate analyses rests not so much on the advantage of taking the average of two determinations as it does on the confidence given by the agreement between duplicates that no gross error has been committed. Past experience provides the criterion for an acceptable agreement between duplicates. There is no way to detect a suspiciously large difference between two measurements except in terms of previous results. It is possible to pass judgment upon the most divergent one of three measurements. It can be predicted that, on the average, a set of three measurements will, once in 20 times, be so distributed that the difference between the two in best agreement will be as little as 3 per cent of the spread between the smallest and largest of the three measurements. For example, the measurements 5.40, 5.41, and 5.72 represent an extreme in uneven spacing that would be unlikely to be met with oftener than once in 20 times if *all* three measurements are free from gross errors. This does not appear to be a very useful rule. Most experimenters would suspect the outlying value even if it had not been as divergent as here taken. This judgment usually rests on some approximate notion of the agreement to be expected, and this information, of course, has not been used by the statistician in arriving at this statistical rule. If the statistician is told that the average difference between duplicates is in the neighborhood of 0.05, he will agree immediately that something is wrong with the high result.

Much of the reluctance of scientific workers to inquire further into the methods of statistics is due to the inability of the statistician, in instances like those discussed above, to do as well as the experimenter in judging the meaning of the data. It is not a fair contest because the researcher usually possesses some familiarity with his methods of measurements, and this gives him some basis, however rough, for making judgments. If the statistician is given access to only a portion of the records of earlier work, he is often able to make sharper judgments than the experimenter. The experimenter, as a rule, fails to make the most of the available information. He is lulled by hollow victories over statisticians lacking any of the information and fails to see that he is not making the most of his data.

It may also be pointed out that in any new line of research the methods of measurement may be new and the reproducibility of the measurements quite unknown at the start. Here experimenter and statistician start even in the available information, and it is here that those equipped with modern statistical tools inevitably possess a marked advantage in passing judgment upon the data, especially at early stages of the work. Without these statistical tools good scientific conclusions can be drawn but usually at needless expense of time and materials.

There is a favorite argument against the employment of statistical tests. It is correctly pointed out that all these tests *assume* that the measurements conform to a particular mathematical law. It is then stated that it takes a very large number of measurements to establish that they are acceptably distributed. This is also true. Triumphantly it is then driven home that very rarely are these large numbers available, so that the whole structure is a house of cards resting upon an unproved assumption. Besides, if the large mass of data were available, no statistics would be needed (presumably however inefficiently they were interpreted!). The answer to this last stand of the conservatives is rather easy. Whenever large sets of scientific measurements have been available, they have shown a truly remarkable similarity in their distribution. It is now known that moderate departures from the assumed or normal distribution exert a very minor and inconsequential influence on the limits given in the tables. The only alternative to using this close approximation is to go on guessing at the answers.

If there is any evidence that the distribution of the measurements departs from the commonly assumed form, there is a simple recourse. Collect the measurements in small sets of three or four and use the averages of these sets as individual observations. The powerful effect of this operation is shown by a simple experiment with a die. There are only six possible results, and for a true die these will have equal frequencies. This distribution has no hump and it is not continuous. Nevertheless the distribution of the averages of four rolls will give a very satisfying approximation to the normal distribution.

Independence of Measurements

There is a requirement that data must meet to make valid the use of statistical techniques. The individual measurements must be independent and on an equal footing. If four samples of a material are prepared for colorimetric analysis and three settings of the colorimeter made on each sample, the twelve readings do not constitute a homogeneous set. Rather there are four sets, and the variation of the three readings within each set furnishes information on the performance of the colorimeter and its user. The four averages associated with the four samples also constitute a set which, with proper allowance for the contribution made by the colorimeter, reveals the variation associated with the analytical process. After all, there are only four samples. Considerable care must be exercised when the measurements fall into groups that are the result of the experimental technique. The discussion of these situations will be taken up in subsequent chapters.

Constant Errors

There is still another objection that may be raised in connection with the use of statistical techniques. It is connected with the concepts of *precision* and *accuracy*. So far all the discussion has been restricted to ascertaining the precision of the measurements. It may be asked what the use is of worrying about precision when all the measurements may be subject to some constant error or bias. Statistical techniques can neither detect nor evaluate such errors, and these may be larger than the variation exhibited by the measurements.

Recall how scientists have become aware of constant errors. They are usually revealed when two or more quite different experimental procedures are employed to measure the same quantity. In pronounced cases each procedure yields a set of closely clustered measurements, but the averages for the procedures are unmistakably separated. What is meant is that the averages disagree more than can be accounted for, considering the precision of the measurements. The measurement of precision is, therefore, an indispensable step in the detection of inaccuracy. The detection and elimination of inaccuracy are experimental tasks—not statistical ones. Statistical techniques, given the necessary experiments, can assist greatly in deciding whether or not inaccuracies exist and whether modification in the procedures for making the measurement has reduced the inaccuracy.

Fortunately in the great majority of experimental programs the specter of inaccuracy is not present. This is due to the skill of the worker in the conduct of his investigation. Thus, in using a permanganate titration to determine iron in an ore, he will standardize, or at least verify, his reagent by an analysis of a known material. The determination of absolute values involves the most painstaking care to achieve accuracy. The determination of differences is quite another matter. There may be a bias in the measurement. With reasonable care this bias will be the same for both items under comparison and vanishes when the difference is taken. The existence of national laboratories with the duty of providing reference standards is recognition of the fact that questions of accuracy involve the utmost refinement of experimental procedures. The great majority of physical and chemical measurements made in research laboratories involve comparisons directly or indirectly with standards.

Rounding Off Data

Finally, in order to determine the precision of the measurements, it is necessary to retain in the data the terminal decimal places which are often dropped. It is a generally followed practice not to give more decimal places than will be meaningful in any use made of the results. This often leads investigators to round off the data and thus suppress a large part of the information originally available for determining the precision. The extra decimal places that are of no consequence to the user are obviously just the ones that do contain the information on the variation between determinations. These extra decimal places should be reported so that the precision may be determined.

CHAPTER 2

The Measurement of Precision

The Average Deviation

The average deviation is often used as a measure of the precision of a set of measurements. It is found by taking the difference between each measurement and the average for the set and then taking the average of these differences without regard to sign. If the measurements are
$$x_1, x_2, \cdots, x_n$$
then
$$\text{Average} = \bar{x} = \frac{x_1 + x_2 + \cdots + x_n}{n}$$
and
$$\text{Average deviation} = \frac{|x_1 - \bar{x}| + |x_2 - \bar{x}| + \cdots + |x_n - \bar{x}|}{n}$$

The use of this measure of precision is so widespread in published scientific work that it is necessary to explain why it is not used in this book. There are several reasons why modern statistics makes almost no use of the average deviation. Two of them will be discussed here.

It is necessary to picture a very large set of measurements of the same quantity, and to perceive that, if some method of computing a measure of precision is adopted, it could be applied to this large set. If this large set of data is arbitrarily divided into many small sub-sets, the same procedure for computing the precision may be applied to every one of the sub-sets. It is inherent in the nature of data that the collection of estimates of the precision obtained from these small sets will themselves be scattered around the average of the estimates. It is obviously desirable to devise a measure of the precision such that, when applied to the above small sets, the results show the least possible scatter. This will improve the chance of an estimate of the precision based upon just one sub-set being close to the average for all the sets. This should interest the experimenter because any given set of data is in fact a sub-set of a much larger number of measurements which might have been made. Judged by this criterion, the average deviation is not the best measure of the precision. It does not seem cogent to advance the simplicity and ease of computation in support of this inferior method of measuring the precision of measurements when a lot of hard work has been expended to make the measurements precise.

There is a second argument against the average deviation which is rarely pointed out but which should, by itself, be sufficient to condemn it in the eyes of experimenters. Suppose the average deviation of a very large set of measurements is computed. Now subdivide the data at random into pairs and calculate the average deviation for every pair. The average of these results will not even approximate the value computed for the whole set but will in the long run be only 0.707 as large. If the whole set of data is broken into triads, the average deviation of the triads will be found to have an average value about 0.816 as large as the average deviation associated with the whole set. The larger the size of the sub-sets the more closely will the average of the estimates of the average deviation approximate the average deviation of the whole set. The factors for sub-sets of five and ten are 0.894 and 0.943. Needless to say there is only one group of data, yet the average deviation leads to different estimates of the precision, depending upon the number of measurements in the sub-groups. In most work the data do come in small sets. Indeed it is not uncommon to see an average deviation computed from as few as three measurements. This will tend to a biased estimate of the precision and in the direction of making the measurements appear to have a better precision than they actually have.

Scientists almost always use the average for a set of data in order to obtain the deviations. Instead of the average the median may be used. The median is the middle measurement if all the measurements are arranged in increasing order of magnitude. It is a known fact that a smaller value for the average deviation will invariably be obtained if the deviations are measured from the median instead of any other value. The point is easily illustrated (Table 1).

TABLE 1. THE AVERAGE DEVIATION

	Measurements	Deviations from Average	Deviations from Median
	3.53	0.05	0.03
	3.56	0.02	0.00
	3.64	0.06	0.08
Total	10.73	0.13	0.11
Average	3.58	0.043	0.037
Median	3.56		

This interesting property of the average deviation does not induce scientists to use the median in place of the arithmetic average. There are circumstances in which the median is used to represent a set of data, but these rarely occur in experimental work.

There has long been available a measure of precision which does not suffer from the disadvantages which have been pointed out for the average deviation. This measure, when computed for each of many subsets drawn from a large set of data, shows a minimum scatter for the estimates. It is, in the statistician's terminology, an efficient measure of the precision. Furthermore the average of the estimates has the same expected value no matter whether the sub-sets are pairs, triads, or any larger group. In the long run the average of the estimates for the subsets will tend to the same value as computed for the large set. It is also of interest that this measure is invariably a minimum for any set of data when the deviations are taken from the arithmetical average. The measure of precision which possesses these good features is called the *variance* and is computed by taking the deviations from the arithmetic average of the data, squaring these deviations, and dividing the sum of these squares by one less than the number of measurements in the set.

The Standard Deviation

The variance, as a measure of precision, does not appeal to scientists because, as a result of the squaring operation, the result cannot be immediately applied to set upper and lower limits around the average. It is customary, however, to extract the square root of the variance and obtain a unit in the same scale as the original measurements. The result of this operation is called the *standard deviation*.

For large sets of data the standard deviation is approximately 1.25 times as large as the average deviation. (The theoretical factor is $\sqrt{\pi/2}$.) The relationship is not true for small sets, the most frequent ones in experimental work.

Other advantages are associated with the variance as a measure of precision. Later in this chapter it will be seen that the *comparison* of estimates of precision and the *combination* of estimates from different sets of data are easy in terms of the variance. The variance holds the same place in the thinking of the statistician that the mole holds for the chemist, so it is not surprising to find that statistical tables are usually set up in terms of the variance or its square root, the standard deviation.

Statisticians customarily use μ to designate the mean (their name for the average) and σ to represent the standard deviation of a set containing an infinite number of measurements. Any actual finite set is considered to be a sub-set from the infinite set, and the estimates of the average and standard deviation computed from such a sub-set are distinguished by labelling them m and s. It greatly clarifies statistical thinking to avoid confusing the true value of the average (or standard deviation) for the infinite set and the estimate derived from a finite set.

Normal Law of Error

Examination of a set of measurements shows the individual measurements clustering more or less closely around the average for the set. Measurements that differ by little from the average are more frequent than measurements which differ considerably from the average. Experience based upon the observation of many sets of measurements in every field of science, coupled with certain theoretical considerations, has led to a general rule relating the frequency of occurrence of a measurement to the amount by which it differs from the population average. This rule is known as the normal law of error. The graph in Fig. 1 shows the

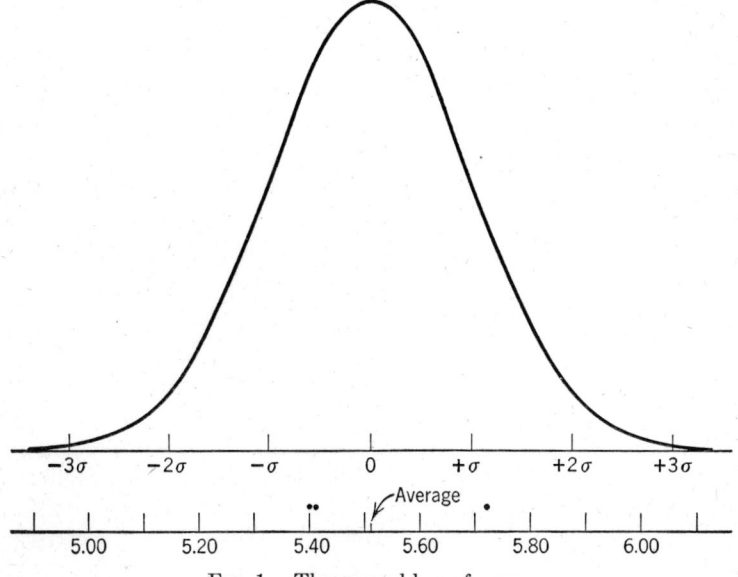

Fig. 1. The normal law of error.

relative frequency of occurrence of measurements when their deviations from the average are expressed in units of the standard deviation. It is a remarkable thing that this single curve approximates so well the scatter of measurements about their average. Two parameters suffice to fit the curve to any particular set. These are the average and the standard deviation. Naturally, the larger the sample the better the estimates of the constants μ and σ. The estimates m and s thus constitute a shorthand account of the information in any particular set of measurements. The normal law of error is so firmly grounded in experience that (in the absence of contrary reasons) it is assumed to apply to sets too small to reveal their approximate conformance. There is a tremendous gain in knowing in advance the shape of the curve. It is this assumption that makes it possible for the statistician to take a set of three measurements (5.40, 5.41, and 5.72) and ascertain whether the calculated values for m (5.51) and s (0.18) are such as to make at all plausible the occurrence of measurements so spaced relative to each other. These measurements have been indicated on the base line under the figure, and it is seen that all three do fall under the main body of the curve.

The points of inflection of this curve are at a distance of one standard deviation from the average. Sixty-eight per cent of the area under the curve is enclosed in the range from minus one standard deviation to plus one standard deviation, which means that, on the average, two out of three measurements will deviate from the average by less than one standard deviation. Plus and minus two standard deviations enclose 95.46 per cent of the area so that about one measurement in twenty may be expected to deviate by more than two standard deviations. It may be shown mathematically that the average difference between duplicate measurements is $2/\sqrt{\pi}$ times the standard deviation. This suggests that it is not necessary to have one large set in order to establish a satisfactory estimate of the standard deviation of the measurements. A sequence of small sets, even duplicates, can be used. In practical work sets of large size seldom occur, whereas small sets (such as duplicate analyses on different lots of material) are usually readily available.

Calculation of Standard Deviation

The advantage of the standard deviation, more particularly its square or variance, is nowhere more clearly apparent than in pooling the information furnished by a series of short sets which may not all be of the same size. The arithmetical operations are first indicated symbolically and then illustrated on some actual measurements. Although the different sets may, and usually do, have in truth different averages, it is presumed that the same precision exists for all the sets. A little reflection shows that the average difference between duplicates (at least over the working range of the process) is independent of the actual magnitude of the measurements. This is generalized for sets larger than duplicates and makes possible the pooling of information from several sets (Table 2).

Table 2. Pooling Information from Several Sets

Set	Measurements	Calculate	Divisor
W	w_1, w_2	$(w_1 - \bar{w})^2 + (w_2 - \bar{w})^2$	1
X	x_1, x_2, x_3	$(x_1 - \bar{x})^2 + (x_2 - \bar{x})^2 + (x_3 - \bar{x})^2$	2
Y	y_1, y_2, y_3, y_4	$(y_1 - \bar{y})^2 + \cdots + (y_4 - \bar{y})^2$	3
Z	$z_1, z_2, z_3, \cdots z_n$	$(z_1 - \bar{z})^2 + \cdots + (z_n - \bar{z})^2$	$n - 1$
		Total of above sums = "sum of squares"	Total = "degrees of freedom"

The last operation to obtain the estimate of the standard deviation is to divide the "sum of squares" by the "degrees of freedom" and take the square root of the quotient. If set Z had contained six measurements, the "degrees of freedom" would have been eleven. The standard deviation has now been estimated with the same assurance that a set of eleven plus one or twelve measurements would provide. In fact fifteen measurements are available, but these are spread over four different lots and lose some of their capacity to furnish information about the precision.

Degrees of Freedom

The divisor that figures in the calculation of the variance (one less than the number of measurements) deserves a word of explanation. It is sometimes called the number of degrees of freedom in the sample. The term "degrees of freedom" has the same meaning in statistics as in geometry and mechanics. By way of illustration, consider the following simple example.

Two independent measurements have been made of the same physical quantity. Each of them reflects

(i) the "true" value of the quantity measured;
(ii) the contribution of error.

It can (and will) be said that our sample possesses two degrees of freedom. The numbers of our sample could, if desired, be plotted on a plane and might, presumably, lie anywhere within some portion of the plane.

Suppose, now, that we calculate two numbers which together are wholly equivalent to the two numbers of the sample, the sum and the difference of the two measurements. One of them, the sum, like the original measurements, reflects both the true value and the error. The difference, however, reflects only the contribution of error. Thus, only one of our two degrees of freedom can be used to display the effects of error alone; the other exhibits the true value as well as the error.

When the set of measurements consists of more than two, the same considerations show that one of the degrees of freedom must be allotted the task of exhibiting the true value, whereas the remaining degrees of freedom show only the effects of error.

The rule that "the number of degrees of freedom allotted to error is one less than the number of measurements" does not hold generally. In fact, it applies only when the fixed "true value" is the only systematic effect in the data. Usually in experimental work several "true values" are built into the experiment. Each of them will require a degree of freedom, and only the remaining degrees of freedom can be used to display the effects of error alone.

Clearly the description of the divisor used in calculating a variance as "the number of degrees of freedom in the set of measurements" is incorrect unless it is understood as an abbreviation for "the number of degrees of freedom in the set of measurements available for estimating error."

The number of degrees of freedom, instead of the sample size, is used as divisor in calculating a variance to ensure the property mentioned earlier, that estimates based on small sets have the same average value as those calculated from large sets. This will not be proved here. It is a well-known fact of mathematical statistics.

An example may give some hint of the idea involved. If a chemist desires to ascertain the precision of an analytical procedure, he may prepare a stock solution and run a series of determinations upon aliquots from the stock solution. The stock solution may have been prepared by weighing out on a rough balance enough material to give a concentration somewhere in the desired range. On the other hand the material weighed out may be known to be of very high purity. It may be that by carefully weighing out a sample and diluting to a known volume the concentration will be known with a much smaller error than the error of the analytical procedure.

After the determinations have been made the analyst will, in the case of the solution made up to approximate strength, take the average of the determinations to establish the strength of the solution. Deviations of the individual determinations will be measured from this average. The data have been used to set this average value, and, in a sense, one constant has been fitted to the data before the deviations can be obtained.

For the carefully prepared known solution the deviations may be taken from the known value. If the experimenter knows, or is willing to assume, that the analytical procedure has no bias or constant error, the deviations of the determinations from the known value can be used to estimate the precision. The average of the determinations is not used in the process of getting the deviations. It is no longer true, as in the first case, that by revealing all but one of the deviations the remaining one can be deduced. In this situation, the degrees of freedom, that is, the appropriate divisor for the sum of squares, is equal to the number of determinations.

In practice there is usually considerable hesitation about assuming the absence of constant errors. If a constant error is present, then this error and the random errors associated with the precision are inseparably merged when the deviations are measured from a known value for the solution. To be sure of measuring just the precision, the data are used to establish a value for the solution and the divisor becomes one less than the number of determinations. It is this estimate of the precision which is appropriate to use in a statistical test of any difference between the known value and the average of the determinations.

Most experimenters are aware that in curve fitting a curve may be made to go through all the points if the curve has an equation with as many constants as there are points. When this is done the plotted points deviate not at all from the curve, so the sum of the squares of the deviations will be zero. The divisor, since one degree of freedom is lost for each constant fitted, would be zero. This ratio is indeterminate and this is appropriate, since obviously no information on precision is available if the curve has been made to go through every point.

Example 1. Calculate the precision of the following spectrographic determinations of zinc:

Set	X	Y	Z
Determinations	3.64	3.55	3.31
	3.56	3.45	3.46
	3.53	3.40	3.29
		3.55	3.59
			3.47
			3.54
Total	10.73	13.95	20.66
Number of analyses, n	3	4	6
Sum of squares of data	38.38410	48.66750	71.21240
$(\text{Total})^2/n$	38.37763	48.65062	71.13927
Difference	0.00647	0.01688	0.07313

Sum of squares $= 0.00647 + 0.01688 + 0.07313 = 0.09648$
Degrees of freedom $= (3 - 1) + (4 - 1) + (6 - 1) = 10$
Quotient gives variance $= 0.009648$
Standard deviation $= \sqrt{\text{variance}} = 0.0982$

The standard deviation of a single determination is about 0.0982 per cent. The average difference between duplicates is related to the standard deviation by the following formula:

$$\text{Ave. difference between duplicates} = \frac{2}{\sqrt{\pi}}\sigma = 1.128\sigma$$

The average difference between duplicates is therefore estimated as $1.128 \times 0.0982 = 0.11$ per cent.

It is usual practice to carry an additional place of decimals while making the calculations. The sums of squares for the variance are often obtained by taking the difference between fairly large numbers. The extra decimal place is required to avoid introducing rounding off errors. It should be dropped at the end of the calculations.

It is customary to calculate the sum of squares by the method used in the above example. It can be shown algebraically that

$$(x_1 - \bar{x})^2 + (x_2 - \bar{x})^2 + \cdots + (x_n - \bar{x})^2$$

is equivalent to

$$x_1^2 + x_2^2 + \cdots + x_n^2 - \frac{(x_1 + x_2 + \cdots + x_n)^2}{n}$$

The squares of the measurements may be looked up in a handbook and summed on an adding machine. This avoids slips in taking differences. Furthermore, as the average seldom comes out to a round number, the use of additional decimal places means that the differences often have more figures in them than the original measurements. The data may be made quite manageable by subtracting a constant (often the smallest one of the measurements) from each measurement. The reduced values are then substituted in the above formula.

It sometimes happens that all the sets are duplicates. In this particular case a little algebra shows that the following simplified formula produces the same result. Square the *difference* between each pair of duplicates. Sum these squares, divide by twice the number of sets, and extract the square root. It will be noted that n pairs of duplicates furnish as much information as $(n + 1)$ measurements all in one set.

Given n sets of duplicate measurements,

$$\text{S.D.} = \sqrt{\frac{1}{2n}(\text{sum of } d^2)} \quad (\text{degrees of freedom} = n)$$

where d is the difference between the duplicates and is taken for each of the n pairs.

Example 2. Determine the standard deviation of an analysis from the following ten pairs of duplicate spectrochemical determinations for nickel:

Sample	Duplicates		Difference	(Difference)^2
1	1.94	2.00	0.06	0.0036
2	1.99	1.99	0.00	0.0000
3	1.98	1.95	0.03	0.0009
4	2.03	2.07	0.04	0.0016
5	2.03	2.03	0.00	0.0000
6	1.96	1.98	0.02	0.0004
7	1.95	2.03	0.08	0.0064
8	1.96	2.02	0.06	0.0036
9	1.92	2.01	0.09	0.0081
10	2.00	1.99	0.01	0.0001
Total	19.76	20.07		0.0247

$$\text{S.D.} = \sqrt{\frac{0.0247}{20}} = 0.035$$

In this example the ten samples are in fact all the same material. The standard deviation may therefore be calculated for the whole set of twenty analyses.

Sum of squares of the 20 analyses	79.348300
$\frac{1}{20}$ square of the sum of the 20 analyses	79.321445
Difference	0.026855

$$\text{S.D.} = \sqrt{\frac{0.026855}{19}} = 0.038$$

The standard deviation is a stepping stone to reach a number of objectives of direct interest to the experimenter. First, it is a suitable measure of the precision of a single measurement. Second, it leads directly to an estimate of the precision of an average merely by dividing by $\sqrt{n}$, where n is the number of measurements entering into the average. Third, the precisions of two sets of measurements are compared by taking the ratio of the two variances [variance = (S.D.)^2]. Fourth, the ratio of the difference between the averages of two sets to the standard deviation of this difference makes it possible to pass judgment upon the difference between the averages. These objectives will be discussed in greater detail.

Confidence Limits

The essential character of the standard deviation is revealed by the following model. Imagine a long series of sets of measurements of the same quantity. The number in each set may vary in any manner. Actually the division into sets is arbitrary because the measurements, all being of the same quantity, really constitute one very large set. Lay off on a vertical line a scale covering the range of measurements encountered. Mark the average of *all* the measurements on this scale and draw a horizontal line through this average. Lay off on this horizontal line as many unit intervals as there are sets. Take the average of the first set and plot it appropriately above or below the horizontal line opposite the first interval. Through this point draw a vertical line extending above and below a distance equal to the standard deviation of the set *average* multiplied by a factor which depends on the size of the set. These factors are shown in Table 3.

TABLE 3. FACTORS FOR CONFIDENCE LIMITS, 5 PER CENT LEVEL

Number of measurements in set	2	3	4	5	10	20
Degrees of freedom (number in set less one)	1	2	3	4	9	19
Factor	12.71	4.30	3.18	2.78	2.26	2.09

The standard deviation must be computed from the first set only, and divided by $\sqrt{n_1}$, where n_1 is the number of measurements in the first set. Plot the average of the second set opposite the second mark and lay off the appropriate multiple of its standard deviation of the average as computed for this set. Proceed in this manner with each of the sets. Then examine the vertical lines which have been drawn to see whether or not they intersect the horizontal line drawn through the grand average of all sets. It will be found, if the number of sets be sufficiently large, that nineteen out of twenty of the vertical lines intersect the horizontal line. It is well to reflect upon the meaning of this phenomenon. Pick at random one of the sets. This set is a group of measurements which might have been made in the experimental program. No information is required beyond that furnished by the set itself to plot its average and lay off the appropriate vertical line for *that set*. If that is the only set, the correct position for the horizontal is of course not known. But the statement may be made, and 19 times out of 20 it will be correct, that the horizontal line lies between the extremities of the vertical segment just drawn. Thus a prediction has been made regarding the *confidence limits* within which the average would lie if based on a very large number of measurements.

This is indisputably useful, the more so since, by an appropriate choice of the multiplying factor, the proportion of times the statement will be correct can be fixed at any desired level. It is apparent that as the factor increases so does the length of the vertical segment and in turn the chance of intersection with the horizontal line through the (unknown) grand average. The use of statistical techniques has made it possible to reduce the chance of an erroneous prediction to any level desired by the investigator.

Examination of the factors, Table 4, shows that they are largest for small sets and high probabilities of being correct. This is in complete

TABLE 4. FACTORS FOR CONFIDENCE LIMITS AT DIFFERENT PROBABILITIES

Proportion of Times That Statement Will Be Correct	Multiplying Factors for S.D. of the Average for Sets of					
	2	3	4	5	10	∞
90 per cent	6.31	2.92	2.35	2.13	1.83	1.64
95	12.71	4.30	3.18	2.78	2.26	1.96
99	63.66	9.92	5.84	4.60	3.25	2.58

accord with the experimenter's own feelings. Everyone knows that a reasonable number of measurements must be made to give confidence in the result. Statistical techniques sharpen the perception and provide a standard procedure for reporting what is seen.

It sometimes troubles experimenters that, depending on which set they inspect, the spread of the limits may be quite small or quite large. The assurance can be given that this is inherent in the nature of measurements. The evidence of many trials, as well as rigorous mathematical proof, shows that on the evidence of a single small set it is possible to make statements relating the average of the small set to the average that would have been obtained if a great many more measurements were made. The solution of this problem for small samples is a great landmark in the development of statistical methods. It was accomplished in 1908 by the English chemist W. S. Gosset, who wrote under the name of "Student."

It is interesting to remember that the experimenter, if he has infinite experience, knows the precision in advance. He can then use this value for the standard deviation instead of an estimate based on a small set. If a small set of measurements is made, the average is, of course, still subject to random displacement. The appropriate multiplying factor for the known standard deviation is listed under the symbol for infinity. It is quite striking that further reduction in the factor for sets larger than twenty is quite small. A quite moderate amount of data submitted to efficient statistical evaluation often makes a more convincing case than an inadequate treatment of a larger amount of work. The worker cannot escape responsibility for passing judgment upon his experimental results. He has to balance the risk of occasional errors in judgment against the amount of work. He should use a good "balance," i.e., statistics, for this purpose and not guess at it.

Example 3. The following five measurements: 2.39, 2.31, 2.38, 2.32, 2.27, yield 2.33 and 0.05 as estimates for μ and σ. Then $2.33 \pm 2.78 \times 0.05/\sqrt{5}$ are the 95 per cent confidence limits for the average.

Comparison of Standard Deviations

There are numerous situations which focus attention directly upon the precision of the results. An analytical method may be given a trial with some modification which it is hoped will improve the precision or at least not lower the precision if the modification is one that simplifies the analysis. Sometimes it is useful to compare the work done by two assistants or two pieces of equipment, or the uniformity among lots from two different suppliers. It will be necessary to have a set of measurements for each of the two items being compared. Let the measurements be:

$x_1, x_2, \ldots x_n$ S.D. $= s_x$ $[(n-1)$ degrees of freedom]

$y_1, y_2, \ldots y_m$ S.D. $= s_y$ $[(m-1)$ degrees of freedom]

The values for s_x and s_y will usually be different. The question is whether the observed discrepancy between s_x and s_y would be likely to arise in sets of size m and n if both populations have the same σ. If the discrepancy between the estimates is greater than would be likely to occur if the x's and y's have the same precision, a strong argument exists for believing the precision to be different for the two sets.

The statistical measure used to compare precisions is the ratio of the squares of the standard deviations. The ratio $F = s_x^2/s_y^2$ has been studied, and the limits are known within which the ratio will normally vary if $\sigma_x = \sigma_y$. A number of cases must be clearly distinguished. It may be that there is no basis for deciding in advance which one of the σ's is the larger. In that case the ratio is taken by placing the larger s^2 in the numerator so that the ratio is always greater than 1. If the circumstances are such that one of them, say σ_x, can only be larger, not smaller, than the other, then the estimate of this one *must* be put in the numerator even if it makes the ratio less than 1. Or, if the claim is made that the modification will improve the precision, then the s belonging to the original procedure must go in the numerator. Table 5 shows, for a few sizes of sets, the upper limits for the ratio F which will, on the average, be exceeded only once in 20 times if σ_x does in fact not differ from σ_y. This holds if the s for the numerator has been designated *before* inspection of s_x and s_y. If the observed ratio is less than unity, the claim has not been substantiated. When the data are inspected

TABLE 5. UPPER LIMITS FOR F, 5 PER CENT LEVEL

		\multicolumn{5}{c}{Degrees of Freedom for Numerator}
		3	5	8	12	20
Degrees	3	9.28	9.01	8.84	8.74	8.66
of	5	5.41	5.05	4.82	4.68	4.56
Freedom	8	4.07	3.69	3.44	3.28	3.15
for	12	3.49	3.11	2.85	2.69	2.54
Denominator	20	3.10	2.71	2.45	2.28	2.12

and the ratio of the larger s^2 to the smaller one is taken so that the ratio is necessarily larger than unity, the limits shown in the table are ones that will be exceeded 10 per cent of the time if σ_x and σ_y are not different. Inspection of the upper limits for the ratio F discloses that moderate-size sets will not provide convincing evidence of a difference in precision unless one of the observed standard deviations is about 50 per cent greater than the other. To detect a small difference in precision the estimates of the standard deviations must be based upon rather large numbers of degrees of freedom. It is fortunate, therefore, that the evidence from a number of small sets may be combined in the manner previously indicated. If long series, of equal length, of duplicate analyses are available by each of two methods (operators, or equipment), the computation of the F ratio is very simple. Simply square the differences between the duplicates, sum the squares for each series, and take the ratio of the sums.

The pooling of information about the standard deviation in order to build up a sufficient number of degrees of freedom requires that the several sets be really comparable. It may happen that the various sets have been obtained under diverse circumstances, possibly by different operators or in different laboratories. When the individual estimates of σ are examined, the values of s are usually found to vary widely. This may raise doubt about the wisdom of pooling the information. A statistical test, devised by Bartlett [*Proceedings of the Royal Society*, A, 160: 268–282 (1937)], is available for passing judgment in such cases. Suppose that K sets are available with varying numbers in the sets. The acceptable upper limits for B may be read from a statistical table of chi square, entering the table with $(K - 1)$ degrees of freedom.

THE MEASUREMENT OF PRECISION

Set	s^2	D.F.	D.F. $\times s^2$	D.F. $\times \log_e s^2$	1/D.F.
1	s_1^2	n_1	$n_1 s_1^2$	$n_1 \log_e s_1^2$	$1/n_1$
2	s_2^2	n_2	$n_2 s_2^2$	$n_2 \log_e s_2^2$	$1/n_2$
.	.	.	.	.	.
.	.	.	.	.	.
.	.	.	.	.	.
K	s_k^2	n_k	$n_k s_k^2$	$n_k \log_e s_k^2$	$1/n_k$
Total		n	$\Sigma n_i s_i^2$	$\Sigma n_i \log_e s_i^2$	$\Sigma 1/n_i$

Compute

$$s^2 = \frac{\Sigma n_i s_i^2}{n} \quad \text{and} \quad n \log_e s^2$$

[overall variance]

also

where

$$B = \frac{1}{C}(n \log_e s^2 - \Sigma n_i \log_e s_i^2)$$

$$C = 1 + \frac{\Sigma \frac{1}{n_i} - \frac{1}{n}}{3(K-1)}$$

Example 4. The following microanalytical determinations of carbon in ephedrine hydrochloride are reported by F. W. Power [*Analytical Chemistry*, **11**, 660 (1939)]. Compare the precisions of the two analysts.

Analyst	H	FWP
	59.09	59.51
	59.17	59.75
	59.27	59.61
	59.13	59.60
	59.10	
	59.14	
Average	59.15	59.62
Calculated sum of squares	0.0214	0.0295
Degrees of freedom	5	3
Variance	0.00428	0.00983
Ratio of variances, F	2.30	
Critical value F, 10 per cent level	5.41	

The evidence for a difference in precision between the two workers is not convincing because the F ratio found is well below the critical value which may be exceeded once in 10 times under the hypothesis that there is no difference in precision. The next chapter will show that there is good evidence for a difference between the averages.

COMPARISON OF STANDARD DEVIATIONS

Example 5. Test for homogeneity of variance the three sets of spectrographic determinations of zinc previously used to give a pooled estimate of the precision. See Example 1.

Set	s^2	D.F.	D.F. $\times s^2$	D.F. $\times \log_e s^2$	1/D.F.
X	0.003235	2	0.00647	−11.46746	0.5000
Y	0.005623	3	0.01687	−15.54267	0.3333
Z	0.014626	5	0.07313	−21.12477	0.2000
Total		10	0.09647	−48.13490	1.0333

$$\text{Pooled } s^2 = \frac{0.09647}{10} = 0.009647$$

$$\log_e s^2 = -4.64111$$

$$n \log_e s^2 = -46.4111$$

$$C = 1 + \frac{1.0333 - 0.1000}{3(3-1)} = 1.1556$$

$$B = \frac{1}{1.1556}[-46.4111 - (-48.13490)] = \frac{1.7238}{1.1556} = 1.492$$

The critical 5 per cent level for chi square, degrees of freedom equal to 2, is 5.98. The value found for *B* is well within this upper limit, and it is concluded that the three sets do not show differences in precision.

Critical values for chi square may be obtained from the last line of the *F* table in the Appendix. This is the line given for ∞ degrees of freedom. Select the entry from the column headed with the desired number of degrees of freedom and multiply this entry by the degrees of freedom. The value, 5.98, used above is obtained by multiplying the *F* value for 2 and ∞ degrees of freedom, 2.99, by 2.

CHAPTER 3

The Comparison of Averages

The t Test

The comparison of averages is an everyday occurrence. If statistics did nothing more than point out that there are a number of clearly defined situations, it would make an important contribution. In fact it provides techniques appropriate to various situations.

The investigator may be confronted with the necessity of comparing the averages of just two lots of material or of more than two. If there are more than two, the task becomes more complicated. Suppose that there are six samples and that this test is a "sleeper"—all six bottles are in reality carefully sampled from the same lot. If a value for the critical differences between lots that would occur only once in a hundred times between *two* samples is used as a yardstick, there is one chance in ten that, among the *six* lots, a difference this large will be found. The spread or range of individual measurements is greater for large sets than for small. This also holds for the averages of sets.

The comparison of just two averages, when the information about the precision has to be based upon the individual measurements from which the averages were computed, involves a certain process of reasoning. If, for example, two materials are compared, the analyses have been performed because there is some doubt about the existence of a difference between the two materials under comparison. The question to be answered, when the results are available, is whether these results are likely to have occurred if there is no difference between the materials. If the results are such as could arise fairly frequently when the materials are the same, then the evidence for a difference is not convincing. If the results are of a kind that would happen very rarely if the materials are the same, then there is a strong argument for concluding that the materials are not the same. It is a statistical task to find a measure for the results and to ascertain the chance of occurrence of various values of this measure when there is no difference. The experimenter can then select a critical value of the measure that suits him; one that would be exceeded only once in 20 times, or once in 100 times if he prefers. In effect he takes the attitude that for values of the measure below this critical value he will consider the existence of a difference "not proved." If a value of the measure exceeding the critical value is obtained, he, recognizing that this is not certain proof of a difference, will consider that the evidence for a difference is sufficient to take action on the basis that a difference does exist between the materials. By selecting the critical values with an eye on the associated chance of occurrence, he can provide protection against being wrong in claiming a difference to the extent that the importance and consequences of the decision appear to warrant.

The tentative attitude, therefore, towards the data is that there is no difference between the sets. This is known as the null hypothesis (Fisher). It applies not only to the averages for the sets but also to their standard deviations. The two estimates of σ^2 are in consequence pooled. Let the sets be

and
$$x_1, x_2, \cdots x_n \quad \text{(average equal } \bar{x}\text{)}$$

Then
$$y_1, y_2, \cdots y_m \quad \text{(average equal } \bar{y}\text{)}$$

$$s^2 = \frac{\Sigma x_i^2 - (\Sigma x_i)^2/n + \Sigma y_i^2 - (\Sigma y_i)^2/m}{(n-1) + (m-1)}$$

$$t = \frac{\bar{x} - \bar{y}}{s} \sqrt{\frac{nm}{n+m}} \qquad [(n+m-2) \text{ D. F.}]$$

The estimated standard deviation of the individual measurements is s; that of the difference between the averages is

$$\frac{s}{\sqrt{nm/(n+m)}}$$

The measure t is the ratio of the difference between the averages to the standard deviation of this difference. It was "Student" in his classic paper of 1908 (*Biometrika*, **VI**, 1–25) who first revealed the properties of this measure together with a table showing the chances that various limiting values of t would be exceeded if both sets came from the same source. Some limiting values of t are listed in Table 6.

TABLE 6. VALUES OF t EXCEEDED ONCE IN 100 TIMES

D.F.	2	4	8	12	30	∞
t	9.92	4.60	3.36	3.06	2.75	2.58

NOTE. A modern version of "Student's" table, taken from Fisher and Yates, is given in the Appendix.

Notice that should m equal n, that is, if the two sets have the same number of measurements, the formula for t reduces to

$$t = \frac{\bar{x} - \bar{y}}{s} \sqrt{\frac{n}{2}}$$

Example 6. Use the data in Example 4 to compare the averages found by the two analysts for the per cent of carbon in ephedrine hydrochloride.

Analyst	H	FWP
Number of analyses	6	4
Average per cent carbon	59.15	59.62
Sum of squares	0.0214	0.0295
Degrees of freedom	5	3

$$s^2 = \frac{0.0214 + 0.0295}{5 + 3} = 0.00636$$

$$s = 0.0797$$

$$t = \frac{59.62 - 59.15}{0.0797} \sqrt{\frac{6 \times 4}{6 + 4}}$$

$$t = \frac{0.47}{0.0797} \times 1.549 = 9.13 \qquad \text{(8 D.F.)}$$

Critical values for t, 8 D.F., are:

$$\begin{array}{ll} 5 \text{ per cent} & 2.306 \\ 1 \text{ per cent} & 3.355 \end{array}$$

The value found for t is much larger than the critical 1 per cent value for t. It is concluded that the observed difference, 0.47, in per cent carbon is not reconcilable with the zero difference expected if both analysts have the same bias. The standard deviation of the average difference is

$$\frac{0.0797}{\sqrt{2.4}} = \frac{0.0797}{1.549} = 0.05145$$

The 95 per cent confidence limits for the average difference, 0.47, are $0.47 + 2.306 \times 0.05145$, and $0.47 - 2.306 \times 0.05145$, or 0.59 and 0.35 per cent.

These few analyses have been sufficient to show that the results of analyst H may be expected to be low relative to those of FWP by an amount somewhere between 0.35 and 0.59 per cent.

The average of the six analyses by H is 59.15 with a standard deviation for a single analysis of $\sqrt{0.0214/5} = 0.0654$. The standard deviation of the average of the six analyses is $0.0654/\sqrt{6} = 0.0267$. The difference between the theoretical carbon content (59.55 per cent) and the average of the analyses may be examined by calculating

$$t = \frac{59.55 - 59.15}{0.0267} = \frac{0.40}{0.0267} = 15.0 \qquad (5 \text{ D.F.})$$

This is far larger than the 1 per cent critical value for t (5 D.F.) and may therefore be considered strong evidence of a bias in the analysis or of an impure sample of ephedrine hydrochloride.

The average for the four analyses by FWP is 59.62. The standard deviation of the average is

$$\sqrt{\frac{0.0295}{3 \times 4}} = 0.0496$$

The 95 per cent confidence limits, $59.62 \pm 0.0496 \times 3.182$, or 59.78 and 59.46, bracket the theoretical value 59.55 for the expected carbon present. These four analyses are therefore consistent with the theoretical result.

It may happen that the standard deviation of the individual measurements is known from earlier work. In that event the measure t is computed exactly as before, and the limiting values of t are taken from the bottom line of the table corresponding to infinite degrees of freedom. Before "Student" this was the only line of the table known, and this accounts for the early insistence on "large samples" for the application of statistical procedures. When the degrees of freedom are a dozen or more, the critical values are not seriously larger than those which hold for infinite sets. For small sets, the factors are much larger, showing that, with limited amounts of data, larger differences between the set averages are required if the ratio t is to reach the critical value which is considered evidence of a difference between the set averages. This is precisely in line with common experience. Many workers are reluctant to base any conclusions on small amounts of data. The fact is that a t ratio of 4.60 resulting from the comparison of two sets of triplicates (4 D.F.) is statistically equivalent to a t ratio of 2.75 obtaining from the comparison of two sets of sixteen measurements.

When exploratory investigations are under way, the number of measurements available is often small. As an example consider the case where each of three different materials has been analyzed once by a standard procedure and once by a modification of this procedure.

THE COMPARISON OF AVERAGES

The analytical results go by pairs, one pair for each material. The question is: Do the methods agree? If they do, the differences obtained by subtracting for each material the result by method Y from the result by method X are equally likely to be positive or negative. A fairly short series of such pairs all giving the same sign is generally accepted as good evidence of a disagreement between the methods. If the methods are in complete agreement, the algebraic average of a long series of such differences will approach zero. The question may be restated by asking whether the average of any short series is inconsistent with the zero average expected on the assumption that the methods do not differ. What is needed is the standard deviation of this average difference. This is obtained by the following procedure:

Material	Original Method X	Modified Method Y	Difference	Sums
1	x_1	y_1	$d_1 = x_1 - y_1$	$x_1 + y_1$
2	x_2	y_2	$d_2 = x_2 - y_2$	$x_2 + y_2$
3	x_3	y_3	$d_3 = x_3 - y_3$	$x_3 + y_3$
Average	$\bar{x}$	$\bar{y}$	$\bar{d} = \bar{x} - \bar{y}$	

The statistical measure here is the same t as before.

$$t = \frac{\bar{d} - 0}{s_d}\sqrt{n} = \frac{\bar{d}\sqrt{n}}{s_d} \qquad [(n-1) \text{ D. F.}]$$

where

$$s_d^2 = \frac{1}{n-1}(d_1^2 + d_2^2 + d_3^2 - n\bar{d}^2) \quad \text{and} \quad n = 3$$

the sign of t being taken as positive. Notice that $n\bar{d}^2$ is equal to the square of the total of the d's divided by n.

Example 7. The following percentages of lead were found for three different heats of an alloy. Method X gives results which average 0.19 per cent higher than those obtained by method Y. Is this difference significant; i.e., does the variation among the three differences support the contention that the average difference is in fact zero?

| Heat No. | Per Cent Lead | | Difference | Total |
	Method X	Method Y	$X - Y$	$X + Y$
1	2.68	2.58	0.10	5.26
2	2.55	2.34	0.21	4.89
3	2.29	2.03	0.26	4.32
Average	2.51	2.32	0.19	

$$s_d{}^2 = \frac{1}{3-1}(0.0100 + 0.0441 + 0.0676 - 3 \times 0.0361) \quad (2 \text{ D.F.})$$

$$= \tfrac{1}{2}(0.1217 - 0.1083) = 0.0067$$

$$s_d = 0.082$$

$$t = \frac{0.19\sqrt{3}}{0.082} = 4.01$$

Critical value for t at 5 per cent level, 2 D.F., = 4.30.

The calculated value for t is slightly less than the critical value and therefore it is not quite convincing that method X is characterized by giving higher results than method Y. The result is strongly suggestive and invites further analyses to settle the issue.

It will be observed that s_d is not the standard deviation of either method but of a constructed quantity, namely, the difference between the methods. It is this difference that is under investigation. Presumably, if there is a disagreement between the methods, this difference is independent of the material analyzed when the materials are similar in nature and, in some cases, do not cover too wide a range of the element being studied. The result is that, with no duplicates, with all six measurements characterized individually by being made on one or the other of three compounds by either the X or the Y method, it is, nevertheless, possible to pass judgment on the agreement between the methods. Three differences furnish only two degrees of freedom, so that the critical t value which is acceptable evidence of a disagreement is rather large. However, this is just the situation in which the experimenter is usually able to collect a longer series.

The F Test

Scientists, looking at the preceding half-dozen measurements, will realize that they also furnish information concerning possible differences among the three materials. Statistics is ready with a criterion to judge this question. It depends upon the well-known fact that the sum of two measurements has the same σ as the difference. In any particular instance the estimate s_d, obtained from the differences, will vary from the estimate s_s, calculated from the sums. It is for the moment assumed that the three materials are really identical. On the basis of this assumption the s computed from the sums is another estimate of σ. The comparison of two estimates of the standard deviation has already been discussed.

The F measure of the ratio of two variances provides critical values which will rarely be exceeded *if* the variances are both estimates of the same σ^2. When the critical ratio is exceeded, it implies that the variances are different, that is, that differences exist among the materials. The estimate s_s for the sums goes in the numerator because it is clearly apparent that differences among the materials will tend to spread the three sums more than would be the case if the materials were alike. This has the effect of increasing the variance, and this effect is known in advance. Hence the value for s_s goes on top even if, in individual examples, it is the smaller of the two estimates. Where it is smaller it is merely the luck of the measurements because it is inconceivable that the materials are more alike than being all one material. And if they are all samples of one material, the sums in the long run must have the same average s_s^2 as the s_d^2 obtained from the differences. This means that the expected long-run average of the ratio F is unity. Values less than unity are known to be meaningless. Values much greater than unity have a ready interpretation—the existence of differences among materials.

Example 8. Using the same data as in the last examples, the estimate s_s^2 for the sums is 0.2242, and the ratio F equals 33.5. Both s_d^2 and s_s^2 have 2 D.F. The critical F value, 19.0, is exceeded.

The number of materials may be as large as desired provided that only two methods are used. From symmetrical considerations it is evident that several methods may be compared in this manner if not more than two materials are used. The extension of the above procedure should three or more analytical methods be used on each of several materials will be discussed in Chapter 6.

One other case remains to be studied: the comparison of several materials by one analytical procedure with varying numbers of determinations on each material. The data obtained may appear as in Table 7.

TABLE 7. COMPARISON OF SEVERAL MATERIALS

Results on Each Material

W	X	$\cdots$	Z
w_1	x_1		z_1
w_2	x_2		z_2
$\cdot$	$\cdot$		$\cdot$
$\cdot$	$\cdot$		$\cdot$
$\cdot$			$\cdot$
w_n	x_m		z_p
Σw_i	Σx_i		Σz_i

Again the scrutiny of the averages for possible differences among them rests upon a ratio of variances. Since but one analytical procedure has been employed, it may be assumed that the several sets provide estimates of the same σ. These estimates may be pooled as described in Chapter 2. This pooled estimate, computed from the variation *within* the sets, is indicated by s_w^2. If the materials do not differ, the variation among the averages for the sets is due solely to the standard deviation of the individual measurements. The problem here is to obtain from these averages, or the corresponding totals, a second estimate of σ^2. If the materials are different, this estimate will tend to be larger than s_w^2. The totals for the K different materials yield the second estimate of σ^2 by means of the formula:

$$s_b^2 = \frac{1}{K-1}\left[\frac{(\Sigma w_i)^2}{n} + \frac{(\Sigma x_i)^2}{m} + \cdots + \frac{(\Sigma z_i)^2}{p} - \frac{T^2}{n+m+\cdots+p}\right]$$

where T is the total of all the measurements.

The formula is a little complicated because the several totals are associated with different numbers of determinations. If, as quite often happens, there is the same number, n, of determinations in each of K sets, the n may be factored out, giving

$$s_b^2 = \frac{1}{n(K-1)}\left[(\Sigma w_i)^2 + (\Sigma x_i)^2 + \cdots + (\Sigma z_i)^2 - \frac{T^2}{K}\right]$$

or equivalently

$$s_b^2 = n\frac{1}{K-1}[\bar{w}^2 + \bar{x}^2 + \cdots + \bar{z}^2 - KA^2]$$

where A = grand average.

In the last expression the various averages may, for the moment, be considered a set of observations. Apart from the factor n the expression is an estimate of the variance of these observations. Remember now that these observations are all averages of n measurements and that averages have a variance equal to $1/n$ that of the individual measurements. The factor n is introduced to blow this estimate up to the same basis as the previously obtained estimate, s_w^2, of the individual measurements. Once the estimates s_w^2 and s_b^2 are obtained the rest is easy. The ratio $F = s_b^2/s_w^2$ is interpreted as before, entering the top of the table with $(K-1)$ degrees of freedom and the left-hand margin with $(n+m+\cdots+p-K)$ degrees of freedom.

Example 9. The data given in Example 1 may be used to illustrate the comparison of sets with different numbers of measurements in the sets.

Set	X	Y	Z	$X + Y + Z$
Determinations	3.64	3.55	3.31	
	3.56	3.45	3.46	
	3.53	3.40	3.29	
		3.55	3.59	
			3.47	
			3.54	
Total of determinations	10.73	13.95	20.66	45.34
Average for set	3.58	3.49	3.44	
Square of total	115.1329	194.6025	426.8356	2055.7156
Divisor (no. in set)	3	4	6	13
Quotient	38.37763	48.65062	71.13927	158.13197

$$s_b^2 = \frac{158.16752 - 158.13197}{3 - 1} = \frac{0.03555}{2} = 0.01778 \quad (2 \text{ D.F.})$$

$$s_w^2 \text{ (see Example 1)} = 0.009648 \quad\quad\quad (10 \text{ D.F.})$$

$$F = \frac{0.01778}{0.009648} = 1.84$$

Critical value F, 5 per cent level, is 4.10 (2 and 10 D.F.). The calculated value for F is below the critical value; therefore the agreement between the set averages is acceptable as measured by the agreement between determinations within sets. Additional determinations might detect differences between the sets.

CHAPTER 4

The Resolution of Errors

Sampling and Analytical Errors

It is common knowledge that an analysis is no better than the sample used for the analysis. In consequence the qualifying phrase "on the sample as received" is often included in a report. Not only does it go against the grain to expend fine analytical skill on a doubtful sample but also there is a responsibility to see to it that the sample is worthy of confidence. The analyst should not feel content to have his duplicates on one sample show much better agreement than analyses based on different samples. He should be concerned to determine, for example, whether he can do more for his client by running single determinations upon each of three samples instead of duplicates on each of two samples. This might reduce the amount of analytical work.

Casual inspection often suffices to show that the analytical work is much better than the sampling. Again the use of a quite simple numerical procedure makes it possible to predict the standard deviation to be attached to an average for any schedule which specifies the number of samples taken and the number of analyses on each sample.

The first requirement is to collect data covering a series of lots from each of which at least two samples are available and to have two or more determinations on each sample. The data may be tabulated as shown in Table 8.

TABLE 8. DATA FOR RESOLUTION OF ERRORS

Lot No.	1	2	3	$\cdots$	n
First sample	x_1	x_2	x_3	$\cdots$	x_n
	x_1'	x_2'	x_3'	$\cdots$	x_n'
Total	X_1	X_2	X_3	$\cdots$	X_n
Second sample	y_1	y_2	y_3	$\cdots$	y_n
	y_1'	y_2'	y_3'	$\cdots$	y_n'
Total	Y_1	Y_2	Y_3	$\cdots$	Y_n
Lot total	L_1	L_2	L_3	$\cdots$	L_n

The standard deviation, s_a, of the analytical procedure is easily obtained from the $2n$ pairs of duplicates. The $2n$ differences are squared, summed, and divided by $4n$, and the square root taken of this quotient. The standard deviation of the sampling procedure cannot be obtained so directly. True, there are duplicate samples from each lot. These results, however, include the analytical error. If the samples were large enough to run a great many determinations on each one, the average for a sample could be considered known with great precision because the standard deviation of an average is

$$s_{\text{ave.}} = \frac{s}{\sqrt{N}} \to 0 \text{ as } N \to \infty$$

where N is the number of determinations on each sample. In such a case the sampling standard deviation could be obtained directly from the agreement between the duplicate samples available from each lot. When only a small number of determinations are available for each sample, the results depend upon the analytical error as well as upon the sampling error. The agreement between results from duplicate samples will be less close than would be the case if the standard deviation of the analytical procedure were zero.

Let σ_s represent the true standard deviation for sampling and σ_a the true standard deviation of the analytical process. Then the apparent standard deviation σ_s' for samples is determined by the number, N, of analyses made on each sample.

$$(\sigma_s')^2 = \sigma_s^2 + \frac{\sigma_a^2}{N}$$

The data provide an estimate s_a^2 for σ_a^2 and $(s_s')^2$ for $(\sigma_s')^2$, making it possible to solve for s_s^2, the estimate of σ_s^2.

Once estimates, s_a^2 and s_s^2, have been obtained, they may be combined so as to correspond to any proposed scheme. In general, if K samples are taken and each of these samples used for N determinations, the standard deviation of the average of the KN determinations is given by

$$\sqrt{\frac{s_s^2 + (s_a^2/N)}{K}}$$

Various values of N and K may be substituted in this expression in order to form an opinion as to the most acceptable combination of N and K. One conclusion appears immediately, that is, that for a given number of analyses on a lot the maximum precision is attained by making only one analysis per sample.

This is made clear by rewriting the expression under the square root sign as

$$\frac{s_s^2}{K} + \frac{s_a^2}{KN}$$

In any particular situation σ_s^2 and σ_a^2 are fixed, and the number of analyses, KN, has also been decided upon. Regardless of how KN is factored the second term is unaltered. The combined expression becomes a minimum when K is as large as possible, and this is done by making N equal to 1.

The sampling variance, s_s^2, may be expressed as a multiple of the variance of the analytical procedure. If $s_s^2 = Fs_a^2$, the above expression for the standard deviation of the average of NK determinations becomes $s_a\sqrt{(NF+1)/NK}$.

Table 9 shows, for various values of F, some values taken by

TABLE 9. VALUES FOR $(NF+1)/NK$

Determinations per Lot	Number of Samples K	Determinations per Sample N	Ratio of Sampling Variance to Analytical Variance				
			$\frac{1}{2}$	1	2	3	5
4	2	2	0.50	0.75	1.25	1.75	2.75
3	3	1	0.50	0.67	1.00	1.33	2.00
6	3	2	0.33	0.50	0.83	1.17	1.83
4	4	1	0.38	0.50	0.75	1.00	1.50

$(NF+1)/NK$ as N and K take on some commonly used values.

Examination of the entries shows that, unless the reproducibility of the samples is a good deal better than that of the analytical work, one analysis on each of three samples is better than running duplicates on two samples. Running duplicates on each of three samples is not as good as running a single determination on each of four samples.

Duplicate determinations are often made for the purpose of detecting the occurrence of grossly out-of-line results. The absence of duplicates here is filled by the replicate samples. If an obviously out-of-line result is obtained, either sample or determination may be suspect. It is a personal choice whether to perform another determination on the suspected sample or obtain a new sample.

Example 10. A comprehensive study of the determination of sulfur in coal is reported by S. S. Tomkins [*Analytical Chemistry*, 14, 141 (1942)]. Four different coals were sent to eight different laboratories to be analyzed in triplicate by four different methods. Data by the bomb method will be examined. The two laboratories listed below were, for each coal, drawn at random from the eight reporting laboratories.

Coal No.		a	b	c	d
Lab. No.		A	F	H	B
Triplicates		1.17	1.60	0.67	3.18
		1.18	1.62	0.70	3.18
		1.21	1.61	0.69	3.24
	Average	1.1867	1.6100	0.6867	3.2000
Lab. No.		D	D	E	E
Triplicates		1.25	1.54	0.74	3.13
		1.26	1.55	0.69	3.15
		1.24	1.54	0.71	3.20
	Average	1.2500	1.5433	0.7133	3.1600

Presumably the original samples were carefully mixed so that the samples of a, for example, sent to A and D may be considered identical. The obvious differences between laboratories on the same coal are chargeable to the laboratories and not to the samples. By picking pairs of laboratories at random there is achieved a sampling of laboratories, the eight laboratories being in the same way a sample from the large number doing this type of analysis.

The data will be used to estimate (a) the variance between determinations conducted in the same laboratory and (b) the variance between laboratories.

The variance within laboratories is obtained by pooling the information furnished by each set of triplicates. Thus for coal a, laboratory A, the three determinations give a sum of squares of 0.0008667 with 2 D.F. The pooled sum of squares for the eight sets is 0.0080668 with 16 D.F. The pooled estimate of s_w^2, of the variance within laboratories, is $\frac{1}{16}$ 0.0080668 = 0.0005042.

The estimate of $s'_b{}^2$, the variance between laboratories, is obtained by summing the squares of the four differences between laboratory averages for the same coal and dividing by twice the number of differences, or 8.

The four differences between the averages are

$$0.0633; \quad 0.0667; \quad 0.0266; \quad 0.0400$$

and the sum of the squares of these is 0.01076334. It is, in fact, easier to take the differences between the totals instead of the averages. The differences between totals are

$$0.19; \quad 0.20; \quad 0.08; \quad 0.12$$

and are seen to be just 3 times the differences between the averages.

These numbers are easier to square and sum. The sum of the squares is 0.0969 and should be divided by 9, since each number was 3 times too large and was then squared.

The result of dividing by 9 is 0.0107667. This is correct, the small difference shown by the first result being due to rounding off the averages. The sum of squares 0.01076667 must be divided by 8 and gives for s'^2_b the value 0.00134583 with 4 D.F.

$$s'^2_b = s^2_b + \frac{s_w^2}{3} = 0.00134583$$

$$\frac{s_w^2}{3} = 0.00016806$$

$$s^2_b = 0.00117777$$

The standard deviation of analyses within a laboratory is $\sqrt{0.0005042}$, or about 0.022 per cent. The contribution made by laboratories is $\sqrt{0.0011778}$, or 0.034 per cent. If a large number of laboratories each ran one analysis on the same sample, the expected standard deviation shown by the results is $\sqrt{0.0011778 + 0.0005042} = 0.041$ per cent, or about twice that shown by analyses within a laboratory.

Sampling in Two Stages

The sampling itself may be a two-stage process. An example in point is the determination of the clean wool content in shipments of wool. The wool is delivered in large bales weighing several hundred pounds. A number of bales are taken from the lot and cores withdrawn by means of a boring device. If the bales are very much alike but heterogeneous within the bale (a rather unlikely state of affairs), it is not so important to spread the cores taken over many bales. If there exists considerable variation in the clean wool content of different bales, there is no escaping the need to spread the cores widely. The problem is further complicated by the fact that the heavy bales are often stored in closely packed tiers in a warehouse. The expense of moving the bales is large compared with the cost of taking a core from a bale once it has been uncovered for sampling.

In order to reduce the analytical costs the cores, fifty or more in number, are combined. A recent study of the sampling of wool was made by an A.S.T.M. committee. Individual cores were subjected to analysis to determine the variance of cores taken from the same bale as compared with the variance of cores from different bales.

Example 11. Two cores are drawn from each of ten bales taken from a shipment of wool. Use the following data to estimate the sampling errors. It is assumed that the error in the quantitative estimation of the clean wool content is small in comparison with the sampling variation between cores.

Bale No.	Clean Wool Content (per cent) Duplicate Cores		Average	Difference
1	55.51	54.32	54.915	1.19
2	55.63	55.58	55.605	0.05
3	62.13	56.51	59.320	5.62
4	63.56	59.84	61.700	3.72
5	61.57	62.68	62.125	1.11
6	54.68	57.64	56.160	2.96
7	58.05	59.70	58.875	1.65
8	58.59	56.70	57.645	1.89
9	54.94	55.30	55.120	0.36
10	58.25	56.61	57.430	1.64
	Grand average		57.8895	

$$s_w^2 \text{ (between cores, same bale)} = \frac{\Sigma(\text{diff.})^2}{2 \times 10} = \frac{65.9489}{20} = 3.2974$$

$$s'_b{}^2 \text{ (between bale averages)} = \frac{\Sigma(\text{ave.})^2 - 10 \text{ (grand ave.)}^2}{10 - 1}$$

$$= \frac{33{,}572.4178 - 33{,}511.9421}{9} = 6.7195$$

(Note: The arithmetic would be reduced by subtracting 54.0000 from every average.)

$$s'_b{}^2 = s_b^2 + \frac{s_w^2}{2} = 6.7195$$

$$\frac{s_w^2}{2} = 1.6487 \quad \text{(variance between cores from same bale)}$$

$$s_b^2 = 5.0708 \quad \text{(variance between bales)}$$

The variation between bales is much greater than that shown by cores from the same bale. The standard deviation of the grand average of the twenty cores, taken two apiece from ten bales, is

$$\text{S.D.}_{\text{grand ave.}} = \sqrt{\frac{s_b^2 + s_w^2/2}{10}} = \sqrt{\frac{5.0708 + 1.6487}{10}} = \sqrt{0.67195}$$
$$= 0.8197$$

If one core be taken from each of thirteen bales, the computed standard deviation of the average of the thirteen cores is

$$\text{S.D.}_{\text{ave.}} = \sqrt{\frac{s_b^2 + s_w^2}{13}} = \sqrt{\frac{5.0708 + 3.2974}{13}} = \sqrt{0.6437} = 0.8023$$

Thus, because of the greater variation between bales than between cores from the same bale, the addition of three bales has more than compensated for cutting the cores taken per bale from two to one. This saves one-third of the analyses. This comparison uses the estimates s_w^2 and s_b^2 computed from the above data as the only available ones for σ_w^2 and σ_b^2.

CHAPTER 5

Statistics of the Straight Line

Detection of Constant Errors

Most analytical procedures require that a sample be weighed and that a subsequent weight of a constituent be determined. Then a percentage figure is obtained by taking the ratio of these and multiplying by 100. If $x_1, x_2, \cdots x_n$ are the weights of a series of samples and $y_1, y_2, \cdots y_n$ the corresponding measurements made on these samples, the chemist usually calculates the ratios:

$$r_1 = \frac{y_1}{x_1}\,; \quad r_2 = \frac{y_2}{x_2}\,; \quad \cdots \quad r_n = \frac{y_n}{x_n}$$

The average of these ratios is taken to represent the data.

It is important to notice the effect on these ratios of some constant error or bias, a, in the measurement of y. The ratios then become

$$r_1' = \frac{y_1 + a}{x_1}\,; \quad r_2' = \frac{y_2 + a}{x_2}\,; \quad \cdots \quad r_n' = \frac{y_n + a}{x_n}$$

The effect of a constant error in the measurement of y on the value of the ratio depends upon the magnitudes of x and y. Ordinarily the disturbance in the value of the ratio r is concealed or avoided by making all the samples of approximately the same weight. If the samples do cover an appreciable range of weights, the variation among the several values of the ratio not only reflects the inevitable random errors in y and x but is also dependent upon the weight of the sample.

The several paired values of x and y may be used to plot the analytical results on graph paper as in Fig. 2. If there is no constant error or "blank" in the analytical process the various points should tend to lie closely along a line passing through the origin. If there is a constant error or bias, all the points are displaced upwards (or downwards) the same amount. The points still lie along a straight line which now intercepts the y axis at a point corresponding to the "blank." The *slope* of the line has not been changed by the presence of a constant error or bias in the method. Since the slope is independent of the presence of a bias, it has an advantage over the ratio r. Furthermore the slope of a straight line is nothing other than the change in y for a unit change in x.

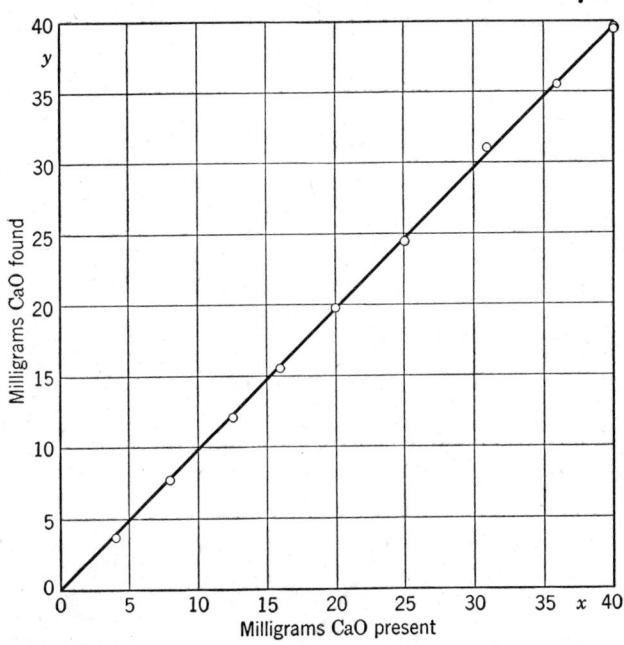

Fig. 2. Graph of analytical determinations.

It is the increment that would be added to the measurement y for an addition of 1 gram to the sample weight. This is precisely what the analyst has set out to determine: the content of y per gram of x.

If the slope of the line is to be used as an alternative means of combining the results of several determinations, there is a real advantage in spreading the sample weights over a reasonable range. If all samples have the same weight, no information is forthcoming about the slope, and the opportunity to detect the presence of a constant error is lost. It is usually thought that a second method, or at least a blank, is required to reveal a bias. This is not the case provided the samples do cover a range in weights.

The least squares formulas for fitting a straight line to a series of points have long been known. The formulas for the constants a and b

STATISTICS OF THE STRAIGHT LINE

in the equation $Y = a + bx$ have been derived on the assumption that the recorded values for x, sample weights in this case, are known exactly. It is sufficient that the errors in x are small compared with the errors in y and that the x's cover an adequate range. This requirement is usually met since sample weights have only the weighing error and the y quantities have all the procedure steps as a source of variation. If all the sample weights are the same, the dispersion of the y values shows the random errors of the analytical procedure. The arithmetical average, $\bar{y} = \Sigma y/n$, is a quantity that makes the sum of the squares of the deviations, $\Sigma(y_i - \bar{y})^2$, a minimum. The formulas

$$a = \bar{y} - b\bar{x}; \quad b = \frac{\Sigma xy - n\bar{x}\bar{y}}{\Sigma x^2 - n\bar{x}^2} = \frac{n\Sigma xy - \Sigma x \Sigma y}{n\Sigma x^2 - (\Sigma x)^2}$$

are similarly designed to minimize the quantity $\Sigma(Y_i - y_i)^2$, where Y_i is the value predicted by the equation $Y = a + bx$ for $x = x_i$, and y_i is the measured quantity for sample weight x_i. The average, $\bar{y}$, is merely replaced by a quantity, $a + bx_i$, to take account of the sample weight, x_i.

Standard Deviation of Slope

It is useful to be able to calculate a standard deviation for the slope b just as was done for the average $\bar{y}$ when all samples had the same weight. Use is made of the quantity $\Sigma(Y_i - y_i)^2$, which is conveniently evaluated by the numerical equivalent

$$(n-2)s^2 = \Sigma y^2 - \frac{(\Sigma y)^2}{n} - \frac{\left(\Sigma xy - \dfrac{\Sigma x \Sigma y}{n}\right)^2}{\Sigma x^2 - \dfrac{(\Sigma x)^2}{n}}$$

The estimate of the variance of a single y measurement is s^2. The sum of squares of deviations of the points from the fitted straight line is divided by $(n - 2)$, not $(n - 1)$, inasmuch as the data have been used to estimate b as well as $\bar{y}$.

The variance for b is equal to

$$\frac{s^2}{\Sigma x^2 - n\bar{x}^2}$$

This approach to the data gives an estimate, b, of the quantity sought, i.e., the proportion of y in x, corrected for any constant errors in the analytical procedure, and also an estimate of the standard deviation for b.

In any given set of data the usual method of taking the average ratio

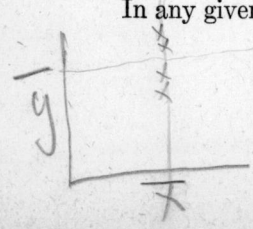

of y/x will not agree with the value for b obtained from the data. It should be noted that the average of the ratios gives equal weight in the average to all determinations regardless of the weight of the sample. This is proper if the analytical errors do increase proportionately with the weight of the sample. The estimate of b assumes that the analytical error is not connected with the sample size over the range of weights employed. In addition the effect of a blank or other contamination which is independent of the weight of the sample is eliminated in the estimate b but not in the average of the ratios.

The slope of a straight line which does pass through the origin is y/x. In such a case it may be expected that the estimate b will be in good agreement with the average of the individual ratios of y/x. The advantage of computing b is that it may reveal a state of affairs that may be overlooked if only the average of the individual ratios is calculated. In addition, the computation for b makes use of information in the data which is not otherwise utilized. When the individual ratios y/x are obtained, the samples are, in effect, all reduced to the same size, and the information available by taking samples of different weights is lost.

The use of the individual measurements and associated sample weights to plot points to which a straight line is then fitted leads to a statistical test to determine whether the line may justifiably be considered to pass through the origin. Merely calculating the average of the ratios *assumes* that the line will pass through the origin.

Standard Deviation of Intercept

The statistical test for passing judgment on the intercept of the straight line is made by calculating the intercept from the formula

$$a = \bar{y} - b\bar{x}$$

and dividing this quantity by the standard deviation of the intercept to get the statistic t. The estimate of the standard deviation of the intercept is given by the formula

$$s_a = s \sqrt{\frac{\Sigma x^2}{n\Sigma x^2 - (\Sigma x)^2}}$$

The value of $t = a/s_a$ is examined by means of the t table, entering the table with $(n - 2)$ degrees of freedom, where n is the number of samples. A value of t in excess of the selected critical value is evidence that the data do not support the expectation that the line goes through the origin. This should lead to further work in the laboratory to determine the cause for this departure.

Example 12. In a study on the determination of calcium in the presence of magnesium Hazel and Eglof [*Analytical Chemistry*, 18, 759 (1946)] obtained the following results with the alcohol method.

x Mg. CaO Taken	y Mg. CaO Found		
4.0	3.7	$\Sigma x^2 =$	6,974.25
8.0	7.8		
12.5	12.1	$\Sigma xy =$	6,884.65
16.0	15.6		
20.0	19.8	$\Sigma y^2 =$	6,796.66
25.0	24.5		
31.0	31.1	$(\Sigma x)^2 =$	54,056.25
36.0	35.5		
40.0	39.4	$\Sigma x \Sigma y =$	53,242.50
40.0	39.5		
Total 232.5	229.0		
Average 23.25	22.90		

$$\text{Slope, } b = \frac{10 \times 6884.65 - 53,242.50}{10 \times 6974.25 - 54,056.25} = \frac{15,604.00}{15,686.25} = 0.994757$$

$$(n-2)s^2 = 6796.66 - \frac{(229.0)^2}{10} - 1552.2181 = 0.3419$$

$$s = \sqrt{0.04274} = 0.207$$

Intercept, $a = 22.90 - 0.994757(23.25) = -0.2281$

$$s_a = 0.207 \sqrt{\frac{6974.25}{15,686.25}} = 0.138$$

$$t = \frac{a}{s_a} = \frac{0.2281}{0.138} = 1.653 \qquad (8 \text{ D.F.})$$

The 5 per cent critical value for t, 8 D.F., is 2.306, and therefore there is insufficient evidence to maintain that the intercept differs from zero more than can be accounted for by the analytical errors.

The slope of the line, 0.994757, falls short of unity by 0.005243. The standard deviation of b is given by the expression

$$s_b = \sqrt{\frac{s^2}{\Sigma x^2 - (\Sigma x)^2/n}} = \sqrt{\frac{0.04274}{1568.625}} = 0.005220$$

$$t = \frac{1-b}{s_b} = \frac{0.005243}{0.005220} = 1.0044 \qquad (8 \text{ D.F.})$$

MEASURING THE FIT OF THE LINE   45

The deviation from unity is little more than the error in b. The slope of the line is acceptably close to unity and the intercept acceptably close to the origin. Nevertheless, Example 14 will show that a line *through* the origin has a slope which departs from unity more than would be anticipated on the basis of the analytical errors.

Measuring the Fit of the Line

There is an alternative equivalent statistical test which may be more understandable. Given a set of data which, when plotted, gives points tending to lie along a straight line, there are available two choices in how a line is fitted. A straight line may be fitted without any consideration as to whether or not the line passes through the origin. Alternatively the best line may be fitted subject to the condition that the line also passes through the origin. It is not difficult to visualize that a line subject to this restriction is not likely to fit the experimental points as well as a line not subject to this restraint. What has to be determined is whether or not the data indicate that a much better fit is obtained for a line free to pass through any intercept.

A measure of the fit is available in the deviations of the plotted points from whichever straight line is fitted. For a line constrained to pass through the origin, that is, $Y = b'x$, the sum of the squares of the deviations of the points from the straight line is given by the formula

$$S' = \Sigma y^2 - \frac{(\Sigma xy)^2}{\Sigma x^2}$$

For a line $Y = a + bx$, free to have any intercept a, the sum of the squares of the deviations of the points from this line is

$$S = \Sigma y^2 - \frac{(\Sigma y)^2}{n} - \frac{(\Sigma xy - \bar{x}\Sigma y)^2}{\Sigma x^2 - \bar{x}\Sigma x}$$

If there are n points, the sum of squares S' is associated with $(n-1)$ degrees of freedom and the sum of squares S with $(n-2)$ degrees of freedom. The latter sum is never greater than S'. The relation between S and S' is shown in Table 10.

TABLE 10. RELATION OF S TO S'

	D.F.	S.S.	M.S.
For line with intercept a	$n-2$	S	$S/(n-2)$
Reduction in S' due to a	1	$S' - S$	$S' - S$
For line through origin	$n-1$	S'	

The F ratio for these two mean squares is

$$F = \frac{S' - S}{S/(n-2)}$$

with 1 and $(n-2)$ degrees of freedom for entering the F table.

A line through the origin gives a variance of $S'/(n-1)$. If the introduction of another constant into the equation improves the fit, it will give a smaller variance $S/(n-2)$. The measure of the improvement is best made by noting how large the quantity $(S' - S)$ is in comparison with the mean square $S/(n-2)$. If the one degree of freedom associated with $(S' - S)$ takes out considerably more than its pro rata share of S', it is evidence for the rejection of the hypothesis that the line goes through the origin.

Example 13. These relationships may be illustrated with the data from Example 12.

$$S' = \Sigma y^2 - \frac{(\Sigma xy)^2}{\Sigma x^2} = 6796.66 - \frac{(6884.65)^2}{6974.25} = 0.4589$$

$$S = (n-2)s^2 = 0.3419$$

	D.F.	S.S.	M.S.
For line with intercept	8	0.3419	0.0427
Reduction due to intercept	1	0.1170	0.1170
Line fitted through origin	9	0.4589	0.0510

$$F = \frac{0.1170}{0.0427} = 2.74$$

The 5 per cent critical value for F, with 1 and 8 D.F., is 5.32, and the result confirms the conclusion drawn in Example 12.

The use of this statistical technique has the further advantage that provision can be made for the systematic examination of the data in case two or more determinations are available for each value of x. The duplicates provide a direct estimate of the variance for the analytical procedure, and it is always desirable to compare this with the estimate of the variance based upon the deviations of the points from the fitted straight line. If there are duplicate determinations for each of n values of x, the tabulation takes on the form:

	D.F.	S.S.	M.S.
Reduction due to intercept	1	$S' - S$	$S' - S$
Line with intercept a	$n - 2$	S	$S/(n-2)$
Line through origin	$n - 1$	S'	
Duplicates	n	D	D/n
Total	$2n - 1$	$S - D$	

If it turns out that $S/(n-2)$ is larger than D/n, significantly so as judged by their F ratio, an unsatisfactory state of affairs is revealed. Either the duplicates have been run in parallel so that their agreement gives an unrealistic estimate of the precision—the value $S/(n-2)$ probably being more nearly a measure of the precision—or else the points do not in fact lie along a straight line. That is, the deviations from the hypothesis of a straight line exceed those to be expected, considering the precision as being properly appraised by the duplicates.

The choice between these two possible explanations is guided by the manner in which the averages for the y's scatter about the straight line. If they are first on one side of the line and then the other in a random manner with only short runs on the same side, the first explanation is the more likely one. If instead the points tend to stay on one side of the line for a series and then drift to the other side and return again to the first side, there is a suggestion that the relationship between y and x is not linear.

The obvious step to take in this event is to fit a curve of higher degree, $Y = a + bx + cx^2$, calculating one more constant, c, from the data. This uses up another degree of freedom, and the mean square associated with c is judged by the residual mean square based on $(n-3)$ degrees of freedom. The arithmetical calculations to determine the constant c and the requisite mean squares for the analysis of variance are very greatly lightened if the x's are spaced at equal intervals and there are the same number of y values for each x. Statistical texts give formulas for second- and higher-degree equations.

Comparison of Slopes

The calcium analyses discussed above are typical of a frequently arising situation where the samples analyzed are known or have been run by a precise reference method. In such cases there exists the anticipation that the line will have a slope of 1 and pass through the origin. If it turns out that the intercept is not significant or, in other words, that the reduction in the error variance due to the constant a is unimportant, there is justification for fitting the line $y = b'x$. The formulas for b' and its standard deviation are

$$b' = \frac{\Sigma xy}{\Sigma x^2} \;;\quad s_{b'} = \sqrt{\frac{s^2}{\Sigma x^2}}$$

where

$$s^2 = \frac{1}{n-1}\left[\Sigma y^2 - \frac{(\Sigma xy)^2}{\Sigma x^2}\right] \qquad [(n-1)\text{ D.F.}]$$

Any difference between b' and unity is examined by calculating

$$t = \frac{1 - b'}{s_{b'}} \qquad [(n - 1) \text{ D.F.}]$$

Example 14. The calcium analyses discussed in Examples 12 and 13 led to the conclusion that the intercept did not differ significantly from zero. That is, the data do not warrant the use of a correction based on a blank determination. Fit a line through the origin and test the agreement of the slope of this line with unity.

$$b' = \frac{\Sigma xy}{\Sigma x^2} = \frac{6884.65}{6974.25} = 0.98715$$

$$s_{b'} = \sqrt{\frac{s^2}{\Sigma x^2}} = \sqrt{\frac{0.0510}{6974.25}} = 0.00270$$

$$t = \frac{1 - 0.98715}{0.00270} = 4.76 \qquad (9 \text{ D.F.})$$

This value for t exceeds the critical value of 3.25 and constitutes evidence that the slope of the experimental line differs from unity by more than a reasonable multiple of the standard deviation of the slope.

Along with the analyses given above, a second set of analyses on the same known solutions was run by a modification of the analytical procedure. The results obtained were

Taken	4.0	8.0	12.5	16.0	20.0	25.0	31.0	36.0	40.0	40.0
Found	3.9	8.1	12.4	16.0	19.8	25.0	31.1	35.8	40.1	40.1

Use these data to calculate

$$b' = 0.99986$$

$$s_{b'} = 0.00149$$

$$t = \frac{0.00014}{0.00149} = 0.09 \qquad (9 \text{ D.F.})$$

The very small value for t indicates that the slope of the line is acceptably close to 1.

The difference between the slopes of two lines may be tested by pooling the estimate of the variance available from each line. For lines of the form $Y = a + bx$

$$s_p^2 = \frac{S_1 + S_2}{n_1 + n_2 - 4}$$

and the variance of the difference between the slopes is given by

$$s_p^2 \left(\frac{1}{\Sigma x_1^2 - \bar{x}_1 \Sigma x_1} + \frac{1}{\Sigma x_2^2 - \bar{x}_2 \Sigma x_2} \right)$$

The value for t is obtained by dividing the difference between the slopes by the square root of the variance of the difference with degrees of freedom equal to $(n_1 + n_2 - 4)$.

For lines passing through the origin

$$s_{p'}^2 = \frac{S_1' + S_2'}{n_1 + n_2 - 2}$$

and the variance of the difference between the slopes is given by

$$s_{p'}^2 \left(\frac{1}{\Sigma x_1^2} + \frac{1}{\Sigma x_2^2} \right)$$

The value for t has $(n_1 + n_2 - 2)$ degrees of freedom.

A comparison of the slopes of the lines through the origin for the old and modified methods for calcium gives

$$s_{p'}^2 = \frac{0.4589 + 0.1399}{18} = 0.03327$$

Variance for the difference between the slopes equals

$$0.03327 \left(\frac{1}{6974.25} + \frac{1}{6974.25} \right) = 0.000009541$$

$$t = \frac{0.99986 - 0.98715}{0.003089} = 4.11$$

The critical value for t, 1 per cent level, 18 D.F., is 2.88, so a difference in slopes may be considered demonstrated.

CHAPTER 6

The Analysis of Variance

Relation between t and F Tests

In Chapter 3 it was seen that the problem of comparing averages has a number of different settings. The different statistical techniques described in that chapter, as well as those in Chapters 4 and 5, are all special cases of a very powerful method for the interpretation of experimental data. This method is called the analysis of variance and was first made available by R. A. Fisher in his book *Statistical Methods for Research Workers*, published in 1925.

The analysis of variance is now a thoroughly proved tool. The first applications were in agriculture and biology, perhaps because the plight of workers in these fields was desperate. At least the technique got a trial, and those who tried it continued to use it. It worked, even if the manner of its working was obscure and imperfectly understood by many who employed it.

It is not difficult to follow the logic which is the basis for the t test. In the previously discussed example of two analytical procedures which were compared on each of three materials the numerical operations seem quite plausible. Should one of the analytical procedures give higher results than the other, this will be reflected in the individual differences and in the average difference. The average difference will be judged by the regularity of the individual differences. The standard deviation of the differences is a measure of the regularity of these differences. When the several differences agree closely in magnitude, confidence is felt in the average of these differences. If the average difference is several times as large as its standard deviation, the existence of a difference between the analytical procedures is considered established. The expression for t

$$t = \frac{\bar{d}}{s_d} \sqrt{n}$$

where $\bar{d}$ = average difference, s_d = standard deviation of individual difference, and $n=$ number of differences, visibly symbolizes the above reasoning. The quantity t is a measure of confidence in the result. The larger the observed difference, the smaller the standard deviation, and the more measurements that are available the greater will be the weight attached to the conclusion that there is a difference between the methods.

If t serves as a suitable measure, so would t^2. The critical values which served for t would merely be replaced by their squares. When the expression for t is squared, the numerator becomes $n\bar{d}^2$. Since $\bar{d} = (\bar{x} - \bar{y})$,

$$n\bar{d}^2 = n(\bar{x} - \bar{y})^2$$

$$= 2n\left[\bar{x}^2 + \bar{y}^2 - \frac{(\bar{x}+\bar{y})^2}{2}\right]$$

The expression in brackets is seen to be the formula for calculating the variance from a set of two values, $\bar{x}$ and $\bar{y}$. The averages $\bar{x}$ and $\bar{y}$ have $1/n$ the variance of individual measurements, and the factor n allows for this. The variance of a difference of two measurements is twice the variance of individual measurements. This shows that $n\bar{d}^2$ is the variance of a difference of two measurements.

In the denominator for t^2 there is $s_d{}^2$, the variance calculated from the three individual differences. Therefore t^2 is in fact a ratio of two variances and is the same as the ratio F. The critical values in the statistical tables for t and F confirm that $t^2 = F$ when *one* degree of freedom is associated with the numerator. It is easy to take the difference between two averages, and the ratio t is designed to take advantage of this special case. The more general case involves the comparison of several averages and requires the F ratio, which for two averages is precisely equivalent to t^2. Once this fact is recognized the use of the F measure loses some of its strangeness.

Return to the six measurements obtained by using two methods of analysis on each of three materials (Examples 7 and 8). Three estimates of variance have now been obtained. They are shown, together with their degrees of freedom, in Table 11. All these estimates refer to

TABLE 11. ESTIMATES OF VARIANCE

D.F.	Source	s^2	$\tfrac{1}{2}s^2$	D.F. $\times s^2/2$
2	From the three differences	$s_d{}^2 = 0.0067$	0.00335	0.0067
2	From the three sums	$s_s{}^2 = 0.2242$	0.11210	0.2242
1	From the two averages	$n\bar{d}^2 = 0.1083$	0.05415	0.0542
5			Total	0.2851

differences or to sums of two measurements, and therefore the factor 2 is included. The next to last column shows the estimates reduced to the bases of individual measurements. The last column shows them multiplied by their degrees of freedom. Now ignore altogether the existence of the two methods and three materials and consider all the measurements a set of six. Compute the variance for this set.

Sum of squares of the six measurements   = 35.1819
Subtract $\frac{1}{6}$ (the sum of the six measurements)^2 = 34.8968

Sum of squares   = 0.2851
Degrees of freedom   = 5
Variance   = 0.0570

Observe that the sum of squares is exactly equal to the total for the last column in the tabulation just above. The three entries in this column are also sums of squares which, on division by their degrees of freedom, yield certain estimates of variance. The three sums of squares associated with the differences, sums, and averages are a partition of the total sum of squares of all the individual measurements. All the variation among the six measurements is accounted for by the three categories. The preceding arguments have indicated that, in the event that the three materials are identical, the variance estimate based on the three material sums has the same expected value as that computed for the differences. If also the modification of the analytical procedure in no wise alters the experimental results, the observed difference between the methods also provides a means of estimating variance. Three estimates of variance can be made, all with the same expectation, provided the experimental variables, here methods and materials, are dummy ones, that is, without effect. If there is an effect associated with one of these variables, say that the modification of the analytical procedure gives results consistently lower, then the variance estimate based on the difference found between the averages for the methods is sensitive to this effect. And, what is equally important, the other two estimates of the variance are not affected. Observe that adding a constant amount, a, to the experimental results obtained by the modified procedure adds the quantity a to *all* the material sums and that this does not alter the estimate of variance computed from these sums. The variance estimate computed from the three individual differences is also independent of the addition of a constant quantity to all the differences.

Use of Variance to Compare Averages

Consider the opportunity the analysis of variance affords the experimenter to detect experimental effects. The possible effect of a modification in the analytical procedure, and of one and two recrystallizations respectively on the material under analysis, may be under study. If none of these variables produces an effect, then the six results are merely six replicate measurements by what in effect is one method on one material. How may we judge whether this is the case? Make the three separate estimates of the variance associated with the effect of method, with the effect of recrystallization, and with the individual differences. The last estimate is not altered by either of the experimental factors and is the basic estimate of the experimental precision. The other two estimates will also have the same expected value for the variance if there is no effect from the experimental variables. The F measure is available to judge departures from this expected value. When critical values of the F ratio are obtained, the null hypothesis of equal variance is rejected and the alternative conclusion accepted that there is a real effect due to the experimental variable.

The arithmetical operations for conducting an analysis of variance have been systemized to reduce the number work. The data (Table 12) will now be reworked as an example of the usual method of computation.

TABLE 12. DATA FOR COMPUTATION

Material Analyzed	Original Method	Modified Method	Total
Heat 1	2.68	2.58	5.26
Heat 2	2.55	2.34	4.89
Heat 3	2.29	2.03	4.32
Total	7.52	6.95	14.47

TABLE 13. COMPUTATIONS FOR THE ANALYSIS OF VARIANCE

	The Six Measurements	The Three-Material Totals	The Two-Method Totals	The Grand Total
Square and sum	35.1819	70.2421	104.8529	209.3809
Divide by: the number of measurements in each item squared	1	2	3	6
Quotient	35.1819	35.12105	34.95097	34.89682
Subtract: $\dfrac{(\text{grand total})^2}{6}$	34.89682	34.89682	34.89682	
Sum of squares	0.28508	0.22423	0.05415	

54   THE ANALYSIS OF VARIANCE

The remainders given in the last row of Table 13 form the entries in the formal analysis of variance in Table 14.

TABLE 14. ANALYSIS OF VARIANCE

Variance Associated with	Based on D.F.	S.S.	M.S. or Variance	F
Methods	1	0.05415	0.05415	16.16
Materials	2	0.22423	0.11212	33.47
Experimental error	2	0.00670	0.00335	
Whole set of six measurements	5	0.28508		

In the general case it is easier to calculate the sum of squares associated with the whole set of six measurements and get the sum of squares for experimental error by difference. The F ratios are obtained by putting the variance for error in the denominator and the others in turn in the numerator. The action of an experimental variable is in the direction of increasing the variance, and only ratios exceeding unity are given in the statistical table of critical F values. If the estimated variances depend upon very few degrees of freedom, the critical F values required to reject the null hypothesis are rather large. Naturally very small sets provide uncertain estimates of the variances. The critical values rapidly diminish with increases in the degrees of freedom. The merit of this interpretation is that it stands squarely on the data alone, drawing no information from sources outside the data.

From the viewpoint of the experimenter there are some other comments to be made about this scheme of evaluating the experimental data. It is obvious that every one of the analytical results is subject to the inherent errors of the analytical process. Each of these results is also subject to a component which is due to the particular material represented and a further possible component associated with the particular analytical method employed. If m_1, m_2, m_3 represent the true values for the three materials, M_x and M_y stand for the possible bias in each method, and ϵ stands for the analytical error, then the data may be represented as follows:

$$x_1 = m_1 + M_x + \epsilon_1 \qquad y_1 = m_1 + M_y + \epsilon_4$$
$$x_2 = m_2 + M_x + \epsilon_2 \qquad y_2 = m_2 + M_y + \epsilon_5$$
$$x_3 = m_3 + M_x + \epsilon_3 \qquad y_3 = m_3 + M_y + \epsilon_6$$

Examination shows that sums of the type $(x + y)$ all contain the quantity $(M_x + M_y)$ as a constant component. If the materials are identical, $m_1 = m_2 = m_3$, and then these are also a constant component of the sums. In this event the sums are subject to analytical errors only as a source of variation among themselves. If $m_1 \neq m_2 \neq m_3$, then additional variation among the sums is to be expected and has to be detected as existing over and above the variation due to the analytical errors.

Similarly the totals $(x_1 + x_2 + x_3)$ and $(y_1 + y_2 + y_3)$ contain the common component $(m_1 + m_2 + m_3)$. If $M_x = M_y$, both totals also contain $3M_x$ as a constant component and differences between them are ascribable to analytical errors only. If $M_x \neq M_y$, a greater difference will be expected between the two totals and has to be judged in terms of the difference which might have occurred when M_x does equal M_y.

Terms of the type $(x - y)$ are as follows:

$$d_1 = x_1 - y_1 = M_x - M_y + \epsilon_1 - \epsilon_4$$

$$d_2 = x_2 - y_2 = M_x - M_y + \epsilon_2 - \epsilon_5$$

$$d_3 = x_3 - y_3 = M_x - M_y + \epsilon_3 - \epsilon_6$$

It is obvious that all these differences have a common component $(M_x - M_y)$ and that differences in bias between the methods cannot, any more than can differences among the materials, be held responsible for variation among the d's. Experimental error alone is the source of the variance among the d's; hence these differences provide the indispensable estimate of variance for comparison with the other estimates. If the other estimates are sufficiently larger, the implication is plain that the materials do differ and that the methods disagree.

Precautions in Using the Analysis of Variance

All the above description makes a very pretty picture and, in a way, suggests that the interpretation of data is a fairly straightforward and simple task. Unfortunately there are several ways in which trouble can be encountered. One of these troubles is the absence, either intentionally or accidentally, of one or another member of the pair of measurements associated with each material. Inspection will show that these incomplete sets must be abandoned. Otherwise the indispensable equivalence among the totals in respect to their common components is destroyed, and the variation among these no longer a simple matter to evaluate. Sometimes it is still possible to proceed at the cost of considerable numerical labor and with some limitations in the interpretation. Here is an issue often raised between statistician and experimenter. The statistician usually requires the symmetry of complete sets if the beautiful simplicity of the numerical analysis is to be preserved. Unless the one drawing up the experimental program does give thought to this matter, the problems of interpretation may become immensely difficult. It is often possible to devise a program that is not too difficult to interpret and is also satisfactory to the laboratory worker. Experience shows that it is a poor bet to expect this to occur by chance; very often when the work has already been done it is discovered that relatively small changes in the program would have contributed very greatly to the amount of information extractable from the data. The responsibility in this respect is clearly in the hands of the experimenter. In later chapters there will be discussed in some detail the features of experimental programs that should be thought through in advance of starting the work.

A second complication may easily arise and be overlooked. The example just discussed took up the comparison of two analytical procedures. It was stated that one of the methods was a modification of the other. There is an implication here that the methods are not too dissimilar. In particular the question that may be asked is whether analyses run by one method have the same variance (or precision) as analyses run by the other. The general method of the analysis of variance requires this to be true. In the illustrative example which is limited to the comparison of just two methods this limitation does not exist. Sums and differences of paired measurements indisputably have the same expected variance (assuming the materials to be identical). The average difference between the methods is judged by the variance of the individual differences, and this is entirely valid. In a study of three or more analytical methods which differ markedly in precision it is obvious that the comparison of the more precise methods is associated with a smaller variance than comparisons involving data with less precision. There is, in this case, no possibility of postulating a variance that holds for all the measurements, and the estimation of variance, from row totals, column totals, and the residue, is without physical meaning. A kind of pooled estimate results if the arithmetic is performed, but this estimate is not strictly applicable to any particular pair of methods. If there is any indication or reason to believe that the precisions of the several methods are different, one recourse is to split the table into parts, grouping together methods that have approximately the same precision. The several parts of the table are then individually submitted to the analysis of variance.

Sometimes the analytical methods under comparison are quite different in character, and the materials subjected to analysis, instead of being closely similar, may differ in their composition so that possible interfering elements occur in some of the materials. In such a case it can easily happen that the experimentally observed difference between two materials depends upon which analytical procedure is used. For, if one procedure can cope with a certain interfering element and the other procedure fails in the presence of this element, then the procedures will agree for a material if the interfering element is absent and disagree when it is present. Thus the difference between two procedures depends on the material being analyzed. This completely destroys the usefulness of the differences $d_1, d_2, \cdots d_n$ as a means of determining the standard deviation of the measurements.

It would be jumping to conclusions to maintain that the ever-present possibility of the analytical methods responding in individual ways in different situations makes it dangerous to attempt an examination of the data by making an analysis of variance. There are many cases in which the analytical procedures are intended to be used on restricted types of materials, or even on one type. The analyst will be able to draw upon his knowledge of chemistry to guide him in such examples. On the other hand he may be in doubt about the matter and wish to assure himself that the procedures under study do respond in the same manner under a diversity of circumstances. The statistical machinery exists to pass judgment upon the existence of this complication. It is nothing more than requiring some duplicates to provide another estimate of variance by which to judge the estimate formed from $d_1, d_2, \cdots d_n$ (the n differences found between the methods on n materials). The next chapter takes up further discussion of this problem.

The complications that can arise in an experimental program and the statistical techniques for the interpretation of experimental data are quite inseparable. In a very real sense statistical theory has to formulate all the possible ways, both pleasing and contrary, in which the experimental results can turn up. Statistical techniques must then be devised to detect, identify, and pass judgment upon what has turned up. Statistics can do this only if the necessary data are available, and these data will be available only provided that the experimenter, conscious of certain possible complications, has done the necessary experiments. The detection of individualistic performance of the analytical procedures, as mentioned above, requires some duplicate analyses. If the chemist is willing to maintain that the modification introduced into the original analytical procedure will not, in view of the known similar compositions of the materials under analysis, make the comparison of methods in any wise dependent on the material, then there is no need for duplicate analyses. If the chemist wishes to collect evidence on this point, then the statistician can help in indicating what analyses must be run in order that the statistical techniques can be applied. These analyses will be, without any question, what the analyst himself, guided by intuition and experience, would also schedule. The chief discrepancy, if any, in schedules lies in the requirement for inherent symmetry in the allocation of the analyses to make the statistical technique involve only simple straightforward numerical operations.

The intimacy of the statistical interpretation and the programming of the experimental work should, by this time, be apparent. To take a very simple case, suppose that two analytical procedures are being tried. If a series of materials is analyzed by one method on Friday and the same series run through with the second method on the following Monday, it may turn out that the second set of results runs high. A statistical test may give a strong case for claiming a difference between the two sets. Whether this difference reflects a real difference between the methods, or whether a change in temperature (or other environmental factor) affected one of the sets, neither the statistician nor the experimenter has any way of knowing. It may be argued that this is just a badly conceived program and that experienced workers don't do that kind of work.

The example given is a crude instance of an experimental consideration which, in more subtle ways, often escapes the attention of the analyst. It is certainly true that most laboratory workers are unaware that both the detection and the outwitting of such day-to-day effects can be achieved by an appropriate scheduling of the program of measurements. Usually no additional work is required, and the precision of the final comparisons is often markedly enhanced when there are environmental circumstances which influence the result. Most often such effects are claimed not to exist, without, however, any evidence being produced that a search has been made for them. Some examples will be taken up in subsequent discussions of the arrangement of experimental programs.

CHAPTER 7

Interaction between Factors

The Concept of Interaction

The word "interaction" is a technical term in statistical terminology. If the bias in an analytical procedure varies from one material to another, then an interaction exists between procedure and materials. This is not strictly the sense in which chemists use the term "interaction," but there is a similarity in the meanings. If the performance of the procedure is independent of the nature of the material being analyzed, there is no interaction.

Another of the ways in which the phenomenon of interaction may be encountered is in the manufacture of a chemical product. At some stage in the process the temperature may be held fixed for a definite period of time. Changing either the temperature or the length of the period may influence the yield. If the yield under the initial conditions was 80 per cent, it may turn out that a higher temperature for the same period of time raises the yield to 90 per cent. Similarly doubling the time period without altering the temperature may raise the yield to 85 per cent. It would be dangerous to assume that the gains are necessarily additive, resulting in a yield of 95 per cent if both the temperature is raised and the time period doubled. The situation may be represented graphically (Fig. 3) by plotting the yields obtained. If the combined effects of raising the temperature and doubling the time are additive, the two lines will be parallel within the limits imposed by the reproducibility of different batches. Temperature and time then produce their effects independently of each other and there is no interaction. On the other hand, if changing both the temperature and the time produces less (or more) than the sum of the separate effects, the lines will not be parallel. The gain resulting from doubling the time period depends upon the temperature, and in consequence the effects of time and temperature are said to have an interaction.

In any given case small departures from parallelism may be caused, or obscured, by the variation among different batches prepared under

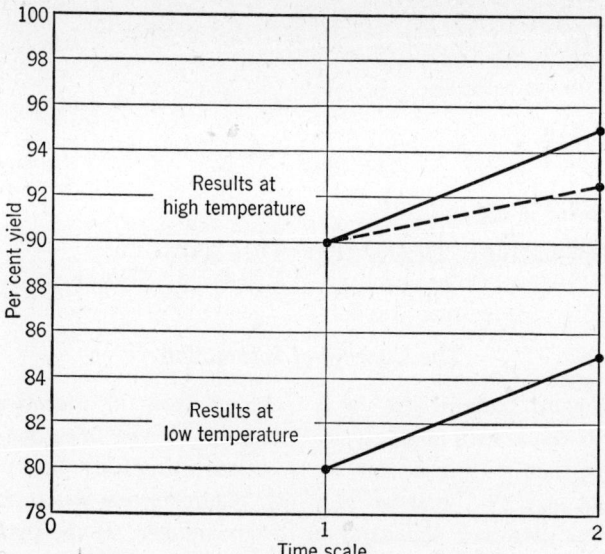

FIG. 3. Solid lines show no interaction between temperature and time. Lower solid line and upper dotted line illustrate interaction between temperature and time.

the same conditions. The problem is to test statistically whether any observed departure from parallelism is greater than could reasonably be expected to occur, taking into consideration the precision of the measurements.

The data may be represented as in Table 15. Triplicate runs are indicated as having been made at each temperature for each time period.

TABLE 15. DATA FOR INTERACTION OF TIME AND TEMPERATURE

Time Period	Low Temperature	High Temperature
1 hour	$y_{11}, y_{12}, y_{13}; \bar{y}_1$	$y'_{11}, y'_{12}, y'_{13}; \bar{y}'_1$
2 hours	$y_{21}, y_{22}, y_{23}; \bar{y}_2$	$y'_{21}, y'_{22}, y'_{23}; \bar{y}'_2$

The gain in yield by raising the temperature for the short period is shown by $(\bar{y}_1' - \bar{y}_1)$, and for the long period by $(\bar{y}_2' - \bar{y}_2)$. For the lines to be parallel these differences should be equal, or equivalently the differences between the differences should be zero. The problem is to test departures of the quantity

$$(\bar{y}_1' - \bar{y}_1) - (\bar{y}_2' - \bar{y}_2)$$

from zero. The expression may be rewritten as

$$(\bar{y}_1' + \bar{y}_2) - (\bar{y}_1 + \bar{y}_2')$$

and, in this form, is seen to be the difference of two sums. The comparison of yields at the two temperatures

$$(\bar{y}_1 + \bar{y}_2) - (\bar{y}_1' + \bar{y}_2')$$

and the comparison of yields for different periods

$$(\bar{y}_1 + \bar{y}_1') - (\bar{y}_2 + \bar{y}_2')$$

are obviously also of the same character. Statistical tests for the comparison of averages have already been discussed. The requisite estimate of the precision is obtained by pooling the evidence furnished by the four sets of triplicates, which make available eight degrees of freedom for the estimate of the variance of these measurements.

Interpretation of Data

The point of view adopted here for looking at these data is closely connected with the usual inspection. Ordinarily, given such a table showing the four averages, the experimenter will compare items in the same row, for each of the two rows, and compare items in the same column, for each column. The increment in yield is noted for an increase in temperature, the time period (top row) being held constant. A similar increment is noted for the second row, and undoubtedly a mental comparison of these two increments is made. This latter comparison is equivalent to contrasting the sum of the entries in one diagonal with the sum of the entries in the other diagonal. It is also numerically the same as comparing the increment in yield produced by doubling the time at the low temperature with the similar increment found at the high temperature. The statistical concept of interaction does correspond to one of the evaluations attempted by the experimenter. It fits into the overall scheme of the analysis of variance and is readily measured in terms of the F ratio.

The argument for employing the statistical approach rests upon the systematic nature of the procedure. Examination of the data leads to one or the other conclusion: an interaction either does or does not exist between the factors over the range studied.

The highway of interpretation forks at this point. In the absence of evidence of interaction it is legitimate to pool the two estimates of the increment in yield produced by the elevation of the temperature. The pooled estimate obtained rests on twice the number of measurements that support the individual increments obtained separately for the short and long time periods. This increases the sensitivity of the experiment in the detection of a temperature effect upon the reaction.

A similar argument holds for the appraisal of the effect of time. Small effects that might otherwise be unconvincing are often established in this manner.

Should the comparison of the diagonal totals indicate that an interaction does exist, the interpretation is that the response of the reaction to a change in temperature depends upon the time period. This means that it is not proper to state a given effect on yield for a given change in temperature unless the time period is also specified. It is an important fact that the statistical test for the presence of interaction uses all the measurements and is not dependent on the comparison of measurements confined to one row or one column.

The existence of an interaction between factors means that the experimental situation is complicated by an interplay of the experimental factors in so far as they influence the quantity observed. The statistical operation for establishing the presence of interactions consists in making the estimate of the variance that is associated with interaction and comparing it with the variance estimate based upon replicate runs. In the absence of genuine interactions both estimates have the same expectation. There are, however, two ways in which the ratio of these variances can exceed unity. First, the presence of an interaction will inflate the variance estimate associated with it. Second, the variance based on replicates may be an underestimate of the real variance of the measurements. This is not a statistical shortcoming but an experimental one.

All too often the duplicate or replicate runs are conducted in parallel on the same day and under the same circumstances. This often results in highly satisfactory agreement among the replicate runs. The data usually will not show such nice agreement if the duplicates are run on separate days. Now the runs which involve varying the experimental factors are apt to be spread over the whole period occupied by the experimental program and, quite apart from the deliberate alterations in the controlling factors, are subjected to a host of environmental variations which may also affect the result. The presence of these is not revealed by parallel duplicate runs. If there is some temporary uncontrolled circumstance, *both* runs respond together and the difference between these results gives no indication of the presence of this disturbance.

The outcome of such a situation is that the precision of the measurements is established under ideal circumstances while the actual experiment is carried out under much less favorable conditions for comparisons. In such a case the real errors in the work are larger than the duplicates indicate them to be. This leads to the danger of judging that real effects have been observed when these results are within the limits that may arise in view of the larger errors that permeate the data. Thus it is possible to have what appears to be evidence of interaction when the variance associated with interaction is compared with the variance computed from duplicates. Naturally this makes for difficulties in interpretation, but statistics has merely revealed this source of trouble and not caused it. Poor experimental planning is the culprit.

Not infrequently it happens that, on the basis of general knowledge, there is no possibility that a real interaction exists between the factors. The analysis of variance, however, may indicate that an interaction exists. The interpretation in this case is obvious. The interaction variance is exhibiting the real variance of the measurements, and the good agreement of the duplicates is due to their being run in parallel. It is such circumstances that lead the statistician to claim with confidence that the experimenter has been overoptimistic about his precision. The argument often gets nowhere because it is not possible to explain quickly how statisticians can estimate the variance by other means than using duplicates, and the experimenter for his part has not in hand the data to show that his duplicates would agree just as well if they had been dispersed through the schedule to the same degree as the various experimental trials.

It is interesting that workers are aware that they may encounter trouble by making replicate runs of one experiment on Friday and comparing these with the runs of another experiment made on Monday. It is sensed that some circumstance may have existed on Monday which affected all results obtained that day. A difference between the averages for the two sets, it is clear, may be ascribed to this difference in circumstances as well as to a possible difference between the experiments. This is exactly the trouble with duplicates run close together. The question may be settled by repeating the first experiment on the second day (instead of running the second experiment). The persistence of a difference between the averages demonstrates that duplicates on separate days do not agree as well as duplicates run in parallel.

There is a simple way to provide protection against the risk that the variance of the measurements will be underestimated when the estimate is based upon the agreement between duplicate runs. All that is necessary is to mix up all the runs in a random order. In the project just discussed the four experiments should be marked on twelve pieces of paper, three pieces of paper being designated for each experiment. The twelve pieces of paper are shuffled and drawn from a hat, and the runs made in the order designated. This avoids any conscious favoring of duplicate runs but admittedly could result in the replicates being done together if a rather unusual draw resulted. Methods of scheduling the experiments to achieve a satisfactory estimate of the variance occupy most of the later chapters. The extraordinary thing is that, by skillful scheduling, the small variance shown by parallel duplicates can be made to hold for the experimental contrasts.

Example 15. It is desired to determine the effect of time of aging on the strength of cement. It is known from previous work that the molds for forming the test specimens are indistinguishable. Two mixes of cement are prepared and six specimens made from each mix. Three specimens from each mix are tested after 3 days and the remaining specimens after 7 days. The idea of using two mixes arises from a desire to be sure that any change in strength due to time is not due to the particular mix tested, or in other words to see whether the change in strength with time is independent of the mix. The same procedure is followed in the preparation of each mix.

It is conceivable that two mixes yield specimens tested at the early date which are in as good agreement as specimens from the same mix. A longer time might give opportunity for a difference between the mixes to manifest itself. In such a case the existence of a difference between the mixes depends on when the test is made. This is precisely equivalent to saying that the effect of the additional period of time is different for the two mixes. When this state of affairs occurs, it is referred to as interaction. If there is sufficient evidence of an interaction, the comparison of the mixes must be done separately for each test time. If there is no convincing evidence of an interaction, the results for the two times may be pooled. This is the efficient way to ascertain whether the difference between mixes is no greater than might be anticipated in view of the variation shown by specimens from the same mix. If the difference between mixes is greater than is expected from the performance of triplicate specimens from the same mix, it is evidence that the technique for repeating the mix introduces some source of variation.

The test specimens are 2-inch cubes and yielded under the indicated loads, which are in units of 10 pounds (Table 16).

TABLE 16. YIELD LOADS FOR CEMENT SPECIMENS

	3-Day Test	Total	7-Day Test	Total	Total
Mix 1	660; 674; 648	1,982	979; 1,038; 1,051	3,068	5,050
Mix 2	661; 624; 652	1,937	1,070; 1,066; 1,053	3,189	5,126
		3,919		6,257	10,176

INTERPRETATION OF DATA

The sum of squares for interaction in Table 18 is obtained by difference, using the results from Table 17. The combined sum of squares for rows and columns is subtracted from that associated with the four totals for the four different experimental conditions. The sum of squares for triplicates is also obtained by subtracting from the sum of squares for

TABLE 17. COMPUTATIONS FOR ANALYSIS OF VARIANCE

	12 Loads	4 Totals of Triplicates	2-Row Totals	2-Column Totals	Grand Total
Square and sum	9,091,732	27,262,638	51,778,376	54,508,610	103,550,976
Divide by	1	3	6	6	12
Quotient	9,091,732	9,087,546	8,629,729.3	9,084,768.3	8,629,248
Subtract	8,629,248	8,629,248	8,629,248.0	8,629,248.0	
Sum of squares	462,484	458,298	481.3	455,520.3	

TABLE 18. ANALYSIS OF VARIANCE

Variance for	D.F.	S.S.	Variance, or M.S.	F
Mixes (rows)	1	481.3	481.3	
Times (columns)	1	455,520.3	455,520.3	198.36
Interaction (time × mix)	1	2,296.4	2,296.4	4.39
Experiments	3	458,298.0		
Triplicate measurements	8	4,186.0	523.25	
All 12 measurements	11	462,484.0		

all twelve measurements the sum of squares for the four experimental conditions. This residue may be checked by calculating for each trial the sum of squares and accumulating over the four sets.

Visual inspection of the data, as well as the very large value for F for the comparison of the 3- and 7-day strengths, shows a tremendous effect from the additional 4 days of aging. It is not easy, by inspection of the raw data, to decide whether the additional 4 days can be considered to have produced the same effect for each mix. The F value for interaction, 4.39, is well below the critical 5 per cent value for one and eight degrees of freedom, 5.32. It cannot be taken that the data give adequate evidence for an interaction. So the conclusion will be that it has not been possible to demonstrate with these data any difference in time effect for the two mixes. This does not rule out the possibility that more data would change the verdict.

As to the question whether specimens from different mixes are as interchangeable as specimens from the same mix, there is also some difficulty in forming conclusions by inspection. The mean square for mixes is, in fact, slightly less than the mean square for specimens within a mix, and so there is no evidence of any difficulty in repeating the mixing procedure.

In summary, whenever the F value for the ratio of the interaction variance to the variance for triplicates falls short of the critical value which has been selected from the tables, the data are considered not to furnish evidence of the existence of an interaction between the experimental factors. This simplifies the interpretation of the experiment. The general effect of time on the strength is shown by the totals (or averages) of all six specimens tested at each time. The effect of mix, at least over the time periods investigated, does not alter the *change* in strength produced by an interval of time. Similarly there is no effect of time on the difference in strength between the two mixes. The row and column totals, or the corresponding averages, are therefore straightforward measures of the effect of time and mix respectively. The evidence furnished by the row or column totals is often designated the "main effect" of the factor involved.

The existence of a main effect will be judged, in the absence of interaction, by the ratio of the variance of the main effect to the variance for triplicates. The pooling of the data, when this is permitted, improves the chance of detecting small effects due to the factors. The presence of an interaction between the factors is a warning that the effect of changing one factor is tied up with the particular value chosen for the other factor. In that case the pooled values, that is, the row and column averages, are apt to be misleading. They represent an average, in the case of the effect of mix, for example, of the results observed at the two times. Nor is it safe to assume that this average effect is what would be found at a time period which was midway between the two time periods actually investigated. It is this state of affairs that makes experimenters balk at taking totals, or averages, of dissimilar things. There is no real difficulty here. The totals, or averages, are used as a means of setting forth the analysis of variance so that the experimenter, by appraising the question of interaction, can determine whether or not he may legitimately *use* the averages. When the evidence indicates that an interaction exists, it also shows that the factors are influencing the result and in a somewhat complicated manner.

The utility of the concept of interaction is not limited to its function as a guide in the pooling of data. There are numerous situations in which the "main effect" of a factor must be judged against the variance associated with the interaction involving the factor.

Consider an industry which is continuously purchasing lots of raw material. The idea may be entertained that the yield of the process will be increased if the raw material is purified by a recrystallization.

Obviously it will not pay to make recrystallizing a standard procedure unless the average gain in yield more than compensates for the cost of this extra step in the process. It is presumed that the lots as received vary in quality so that the advantage of purifying the material varies from lot to lot. Let it be further assumed that the precision of the data is much greater than the variation in gain shown among the lots. The data may be represented in the form shown in Table 19.

TABLE 19. EFFECT OF RECRYSTALLIZATION ON YIELD

Lot No.	1	2	3	$\cdots$	n	Ave.
Yield as received	x_1	x_2	x_3	$\cdots$	x_n	$\bar{x}$
Yield when recrystallized	y_1	y_2	y_3	$\cdots$	y_n	$\bar{y}$
Gain in yield ($y - x$)	d_1	d_2	d_3	$\cdots$	d_n	$\bar{d}$

Experience with a number of lots will be required before it will be possible to venture a prediction concerning future lots. Of course the prediction will not stand up if the source of supply is changed or the supplier changes his product. There is the tacit assumption that the lots already examined are a fair sample from the source of supply.

The confidence in the average, $\bar{d}$, depends upon the variation among the individual d's found for the several lots. Each difference in yield for the crude and purified materials depends upon the lot. There is, in effect, an interaction between the factors, crystallization and lots. The analysis of variance may be set forth as in Table 20 in order to

TABLE 20. ANALYSIS OF VARIANCE

Variance	D.F.	S.S.	M.S.	F
Crystallization	1	$\ldots$	s_C^2	s_C^2/s_I^2
Lots	$n - 1$	$\ldots$	s_L^2	
Interaction	$n - 1$	$\ldots$	s_I^2	
	$2n - 1$	$\ldots$		

examine the effect of recrystallization upon the yield. It would not do to judge the *general* effect of recrystallization by the agreement shown between gains measured by replicate determinations on the same lot. The variance based upon the replicates would be used to judge whether any *particular* lot showed a gain in yield from recrystallizing. The agreement between duplicate measurements in the same lot can throw no light on the variation which may be encountered from lot to lot. Only the interaction variance can show this.

The gain in yield is shown by the comparison of two sets—the crude and purified—and the convenient t test may be used in this case. The

t technique does not draw attention explicitly to the fact that it is the interaction variance that is being used to judge the general effect of recrystallizing. This sometimes leads to an incorrect appraisal of the data when replicate measurements of the gain are available on each lot. If triplicate measurements of d are made on each lot, there are available $3n$ differences from the n lots. These cannot be considered one large set for the estimate of variance, or the $\sqrt{3n}$ used as a divisor to determine the standard deviation of the average difference. Only n lots have been examined. The estimated variance will be much too small if the triplicates agree closely in comparison with the agreement shown between lots.

In the event that a large F is found, which in itself implies that there is an average gain, the next question is whether or not this average gain is at least equal to D, the gain required to compensate for the cost of purification. A little thought shows that, if the true gain δ is just below D in value, the estimate $\bar{d}$ of δ will nevertheless exceed D about half the time. That is, any given series is equally likely to have an average above or below the true value. It is necessary to allow a margin to provide for this contingency. The standard deviation of a single measurement of yield, as determined from the interaction, is s_I. The standard deviation of a difference between two yields is $\sqrt{2}s_I$, and, the standard deviation of the average of n differences is $\sqrt{2}s_I/\sqrt{n}$. If the recrystallization has no effect then, there is 1 chance in 10 of finding a difference as large as $1.64(\sqrt{2}s_I/\sqrt{n})$, the difference being of either sign. The factor 1.64 is used when n is greater than 30. For n less than 30, the factor must be taken from the t table, entering with $(n-1)$. The chance of finding a difference of specified sign is 1 in 20. By requiring $\bar{d}$ to be as large as $D + 1.64(\sqrt{2}s_I/\sqrt{n})$, only once in 20 times will δ, the true value, be less than D. The coefficient for s_I, at the 5 per cent level, for several values of n shows, as would be expected, that as n increases the required excess in $\bar{d}$ over D diminishes.

n	2	4	8	18	32
Coefficient	6.314	1.664	0.948	0.580	0.411

The analysis of variance makes provision for the systematic inspection of all possible interactions when there are more than two experimental factors in the program. Possibly, in addition to the temperature and time factors, the concentration of one of the reactants may be varied. Not only does this introduce possible interactions of concentration with time and with temperature, but, in addition, the nature of the interplay between the effects of time and temperature may be different at the different concentrations.

The effect of temperature may be studied by performing p experiments, one experiment at each of p different temperatures. Time and concentration are held constant. The first set of experiments may be repeated, varying the time period until q different times have been explored. The total number of experiments so far is pq. If the program to date is further repeated for r different concentrations, a total of rpq experiments will have been performed. In the event that all the factors are without influence upon the result, the rpq measurements constitute a set whose variance could be computed with $(rpq-1)$ degrees of freedom. In any case where the factors do influence the result a really satisfactory interpretation calls for ascertaining whether the factors act independently of one another, and the whole program should usually be duplicated to provide an estimate of the variance of a measurement.

The formal analysis of variance for the above program is set down in Table 21.

TABLE 21. ANALYSIS OF VARIANCE FOR THREE FACTORS

Variance	Degrees of Freedom
Temperature	$p-1$
Time	$q-1$
Concentration	$r-1$
Two-factor interactions	
Temp. × time	$pq-p-q+1$
Temp. × conc.	$pr-p-r+1$
Time × conc.	$qr-q-r+1$
Three-factor interaction	
Temp. × time × conc.	$pqr-pr-qr+r-pq+p+q-1$
Total	$pqr-1$
Duplicates	pqr
Total	$2\,pqr-1$

The analysis of variance also serves as a guide in the presentation of the data. If the data indicate the existence of the three-factor interaction, the full array of all the measurements should be shown in a table. If there appears to be no three-factor interaction, the results should be pooled and condensed into three tables (22, 23, 24) showing the relationship for each pair of factors.

The entries in these subsidiary tables will rest upon $2r$, $2q$, and $2p$ experiments and for that reason are more stable and are far more likely to reveal to the eye the relationships involved. If the analysis of variance shows one of the two-factor interactions as not existing, the corresponding subsidiary table may be collapsed into its row and column averages. These averages already appear in the margins of the other two tables; consequently this subsidiary table may be dispensed with entirely. This is merely saying that, if the factors, say temperature and time, act independently, the marginal averages show the effect of time and temperature better than the individual columns and rows because they are based upon more values. If two of the two-factor interactions are non-existent, one subsidiary table for the pair that does show an interaction suffices. This table, together with the overall

TABLE 22. RESULTS POOLED FOR ALL r CONCENTRATIONS

Time	Temperature			
	1	2	...	p
1	–	–	...	–
2	–	–	...	–
.	.	.	...	.
.	.	.	...	.
.	.	.	...	.
q	–	–	...	–

TABLE 23. RESULTS POOLED FOR ALL q TIMES

Conc.	Temperature			
	1	2	...	p
1	–	–	...	–
2	–	–	...	–
.	.	.	...	.
.	.	.	...	.
.	.	.	...	.
r	–	–	...	–

TABLE 24. RESULTS POOLED FOR ALL p TEMPERATURES

Time	Concentration			
	1	2	...	r
1	–	–	...	–
2	–	–	...	–
.	.	.	...	.
.	.	.	...	.
.	.	.	...	.
q	–	–	...	–

averages for the third factor, compactly presents the information in the data. Finally, if there are no interactions whatever, the overall averages for the three factors, taken from the margins of the above tables tell the whole story.

The numerical work is a simple extension of the scheme used for calculating interaction variance for a two-factor experiment. The three subsidiary tables are regarded as two-factor experiments, and the interactions determined for these. The sum of squares for the three-factor interaction is obtained by difference after all the other sums of squares have been calculated.

Interactions of more than three factors are of secondary interest, and even three-factor interactions may not be of much concern. Experience shows that the variance associated with multifactor interaction is very often very little larger than the variance estimated from duplicate measurements. This leads directly in comprehensive programs to the omission of duplicate measurements, the variance associated with multifactor interactions being used in place of the estimate ordinarily provided by duplicates. This makes for economy of effort, especially as an extensive scheme of investigation involves a large number of experiments.

Experimenters are often tempted to run experiments for only certain combinations of the factors. The data in this case cannot easily be examined by the analysis of variance technique—many times not at all.

Inability to apply this statistical technique brings out in the open the difficulty of interpreting such hit and miss programs—a difficulty the experimenter is apparently often unwilling to face, as the data are sometimes merely tabulated and the reader is left "to draw his own conclusions."

There is another advantage in carrying out the complete schedule (or at the least a studied and symmetric selection). Factorial experiments, as they are called, lend themselves exceedingly well to certain groupings or partitions of the whole program. It will be seen later that these groups may be assigned to different workers, or sets of equipment, or days, without introducing into the important comparisons the additional dispersions which would result if these associated environmental factors were not controlled.

CHAPTER 8

Requirements for Data

Planning the Experimental Program

The preceding chapters have dealt with some important statistical techniques which are useful in the interpretation of experimental data. A number of times it was necessary to point out certain pitfalls that may trap the unwary. In every instance these pitfalls were connected with the manner in which the measurements were obtained. In many cases the data themselves can lead to erroneous conclusions, quite apart from the use of formal statistical methods. Indeed, the service that statistical techniques render in bringing out clearly the requirements that data must meet, if they are to be interpreted at all, is fully as important a contribution as is the solution of the mathematical problems involved in the preparation of statistical tables.

It seems a common practice not to worry about the interpretation of the data until after the experiments have been done. Frequently a collection of data is brought to a statistician to see what he can do with it. And often he can do little for the same reason that the experimenter sought help—the objectives of the experimental program were not properly thought out and the data were not collected in such a manner that a valid criterion can be found for passing judgment upon the results.

Replication of Measurements

In order to see more clearly the requirements that data must meet it will be helpful to consider a specific problem. Suppose that the objective is the determination of the effect of an aging period at elevated temperatures on the breaking strength of concrete. In the absence of prior experience it will be necessary to institute a preliminary program. More or less arbitrary decisions have to be made regarding the number of temperatures investigated, the particular temperatures to be chosen, and the number of specimens to be tested. A good deal of work may be saved if the investigator has sufficient experience with the experimental material to make some shrewd guesses. Statistical advice is frequently sought on the number of specimens that should be used. It is not possible to advise on this matter in the absence of a few preliminary measurements and a statement by the investigator of the minimum effect that it is important to detect. It sometimes helps to refer to a table showing, for various standard deviations, the numbers of measurements required to be reasonably sure of revealing effects of specified magnitudes. (See Cochran and Cox, *Experimental Designs*, John Wiley & Sons, Table 2.1.) The immediate circumstances of available time and equipment are often the overriding considerations.

There are one or two other pointers that may also serve as a guide. Eight specimens divided between two temperatures, nine among three temperatures, and twelve allotted to six temperatures, all furnish six degrees of freedom for the estimate of the variance of the measurements. It will assist the examination of the data to test the same number of specimens at each temperature and to space the temperatures at equal intervals on an arithmetic, logarithmic, or other convenient scale. These suggestions do not seem to impose any serious limitation upon the worker.

A simple experiment will serve to illustrate certain essential elements in the program. Suppose that eight specimens are available and that these are divided between two temperatures—a reference temperature and one elevated aging temperature. Four specimens are assigned to each temperature. It is worth while to pause and consider why replicates are run at all. If only one specimen is employed at each temperature, it will be impossible to decide, without drawing on previous experience, whether any observed difference between the breaking loads is due to the aging temperature or is ascribable to the variation in performance encountered between specimens which have been treated alike. Even if there is prior experience to serve as a guide, no protection is afforded against the possibility that one of the specimens had a concealed defect and was not in fact a representative sample. Furthermore, even if the variance is known, the averages of sets of four measurements make it possible to detect an effect one-half as large as an effect that would be considered convincing if only one specimen was tried at each temperature.

Replication serves three purposes: first, to establish from the data an estimate of the variance of the measurements; second, to make possible some protection against faulty specimens and the inclusion of out-of-line measurements; and, third, to gain the benefit of the increased precision of averages over that of individual measurements. All the above is summed up in the usual statement that in order to rely upon the result it must be possible to repeat it.

The emphasis upon the equal division of the specimens is worth noting. The expression for t

$$t = \frac{\bar{x} - \bar{y}}{s} \sqrt{\frac{nm}{n + m}}$$

contains n and m, the number of specimens in the two sets. If $n = m = 4$, the quantity under the square root sign is 2.0. If $n = 2$ and $m = 6$, the quantity becomes 1.5. There are still six degrees of freedom for the estimate s, but the unequal division obviously reduces the value for t, and t has to reach a selected critical value before accepting the conclusion that the elevated temperature has produced an effect. This merely verifies an established belief that it is no use to know one of the quantities very precisely if the other one is subject to large errors.

Arrangement of Experimental Material

There remains one other matter to consider in the experimental procedure. It is so obvious in the simple experiment under discussion that it may seem foolish to dwell upon it. This has to do with the manner of allocating for the reference temperature four specimens out of the available eight specimens, leaving the other four for the elevated temperature. Nevertheless there are at least three methods by which this division can be made. One completely wrong way may be illustrated by the following example. It is believed that impregnating leather with a rubber preparation will increase the wearing quality of the material. The experimenter takes eight pieces of leather and, using his familiarity with leather, *selects* the four best specimens and uses these as controls. The remaining four inferior specimens are impregnated and then found to give more hours of wear on the test machine. The experimenter, feeling that he has leaned over backward, confidently proclaims that he has demonstrated his case. Perhaps he has, but only in a qualitative sense. He has not performed a quantitative experiment in spite of recording the hours of wear for each piece. There is no possible way to ascertain the gain in life resulting from the impregnation with rubber. Presumably it is greater than the figures show, but the increment is hopelessly confused with the differential in wear introduced by his selection, by his manner of allocating the specimens. Deliberate selection of this kind destroys the quantitative character of the study.

There is another kind of selection which frequently improves the experiment. It is based on the sound idea that the two sets should be as alike as possible to begin with. The best two specimens (as far as can be judged) are picked out and one assigned to each set. Then the next best two specimens are taken and divided, and so on until all the specimens are allocated. Two boys on a sand lot choose their teams in a somewhat similar manner. If the experimenter possesses some ability at matching up specimens, this method of selection does balance the two sets off against one another very nicely. An immediate consequence, however, is that the specimens *within* a set have been deliberately made as dissimilar as possible. To use the variance among the specimens constituting a set as a criterion for judging a difference between set averages is a self-defeating process. This variance has, by the conscious choice of the experimenter, been made large. The sets, by being made more alike, are really more sensitive in the detection of an improved effect. The statistical examination makes provision for this by calculating a t test from the differences shown by the matched members of each pair. Alternatively the analysis of variance serves to isolate and set aside the variance between pairs which, by the assignment of the specimens, does not affect the comparison between sets and in consequence must not be included in the estimate of the variance used to judge the difference between sets.

It should be noted that the degrees of freedom for the variance between pairs reduce the number of degrees of freedom available for the estimate of the experimental error. If the circumstances are such that the difference between pairs is negligible, there will be little reduction in the error variance. The reduction in the degrees of freedom available for estimating error without a more than compensating reduction in the error variance is most undesirable. Prior experience with the experimental material is necessary to anticipate and avoid this waste of the degrees of freedom.

Random Arrangements

The remaining way of allocating the specimens to sets is to make no selection whatever. The division of the specimens is left to chance, so that two specimens assigned to different sets have just the same opportunity to be alike as two specimens going to the same set. In this way an observed difference between sets can be evaluated because it is now possible to predict (by observing specimens treated alike) how large this difference could be if the treatment applied to one of the sets is without effect.

It is instructive to examine more closely the implications of the assignment by chance of the specimens to the two sets. To do this it is convenient to postulate that the second set of cement specimens, intended for the higher aging temperature, is aged under the same conditions as the reference set. If the pieces are then broken the eight results may be examined to see what sort of things can turn up when the treatment is known to be without effect (because no treatment was given). It requires no stretch of the imagination to conceive that the eight results, if the loads are read closely enough, will all be different. Granted this, it is clear that one of the possible chance partitions of the specimens ends up with the four highest loads in one set and the four lowest in the other. If the experiment had actually been carried out, that is, one set aged at the higher temperature, and all the values in one set found to be larger than the largest value in the second set, most workers would conclude that the higher aging temperature produced an effect. It is not difficult to calculate the chance that this unique division will be obtained. The number of ways in which four items can be selected from eight different items is $8!/4!4!$, or 70. These seventy sets form thirty-five pairs, since any given set of four immediately fixes the composition of the residual set of four. In consequence there are just thirty-five ways in which the objects may be split into two sets, and one of these divisions is unique in having all the values in one set greater than all four values in the other set.

The experimenter runs this risk of obtaining an experimental outcome which, by mere inspection, would lead him to conclude there was an effect produced by subjecting one set to a higher aging temperature. There is no protection against this eventuality except increasing the number of specimens. Nor does such an increase in the number of specimens guarantee the experimenter against a mistaken conclusion. Even a hundred items can be drawn so that the first fifty drawn are the fifty largest values, although the probability of this taking place is extremely small. What the experimenter has to do is to settle with himself what degree of protection he desires, and then, if he is willing to undertake the amount of work this entails, proceed with the program.

It is inefficient to base one's judgment on the ranking of the values without utilizing the specific values associated with each specimen. It is the contribution of statistical techniques that the specific values can be used to sharpen the discrimination between the sets. To show how this works a set of eight measurements of the breaking strength for eight specimens treated exactly alike will be put through the formal statistical test for every one of the possible thirty-five divisions of the eight values. The observed breaking loads for the eight specimens are as follows:

Specimen	a	b	c	d	e	f	g	h
Psi $\times 10^{-2}$	150	162	163	171	174	176	181	194

One set of each of the thirty-five divisions is given in Table 25, together with the difference in totals between the set and its complement set. The value of t for each of the divisions is also given.

TABLE 25. DIFFERENT DIVISIONS OF EIGHT SPECIMENS

	Diff.	t		Diff.	t		Diff.	t		Diff.	t		Diff.	t
abcd	79	3.16	abdg	43	1.17	abgh	3	0.07	aceh	9	0.22	adfg	15	0.37
abce	73	2.62	abdh	17	0.42	acde	55	1.62	acfg	31	0.80	adfh	11	0.27
abcf	69	2.34	abef	47	1.31	acdf	51	1.46	acfh	5	0.12	adgh	21	0.53
abcg	59	1.79	abeg	37	0.98	acdg	41	1.10	acgh	5	0.12	aefg	9	0.22
abch	33	0.86	abeh	11	0.27	acdh	15	0.37	adef	29	0.74	aefh	17	0.42
abde	57	1.70	abfg	33	0.86	acef	45	1.24	adeg	19	0.47	aegh	27	0.69
abdf	53	1.53	abfh	7	0.17	aceg	35	0.92	adeh	7	0.17	afgh	31	0.80

The important point here is to examine the thirty-five computed values for t obtained on a homogeneous set of measurements. These t values will indicate the performance of the t test when the null hypothesis is true. They reveal how large a critical value for t must be selected as a dividing line in order that larger values, because of their infrequent occurrence, may be considered evidence for the existence of a real effect. The t values have been collected in classes in Table 26, using as class limits critical values taken from the statistical table for t.

TABLE 26. VALUES FOR t FOR DIFFERENT DIVISIONS OF EIGHT SPECIMENS

t Values Taken from t Table	Proportion of t Values Expected in This Range	Number of t Values Expected	t Values Computed	Number of t Values Found
0 –0.265	0.2	7	0.07; 0.12; 0.17; 0.22;	7
0.265–0.553	0.2	7	0.27; 0.37; 0.42; 0.47; 0.53	8
0.553–0.906	0.2	7	0.69; 0.74; 0.80; 0.86	6
0.906–1.440	0.2	7	0.92; 0.98; 1.10; 1.17; 1.24; 1.31	6
1.440–1.943	0.1	3.5	1.46; 1.53; 1.62; 1.70; 1.79	5
1.943–2.447	0.05	1.75	2.34	1
2.447–∞	0.05	1.75	2.62; 3.16	2
Total	1.00	35.00		35

(Values underlined occur twice)

The t values taken from the table have been mathematically established as the limits within which certain percentages of t values may be expected to fall. These percentages have been multiplied by 35 to give the number of t values expected in each bracket. The thirty-five actual values obtained with the above eight measurements have been collected in the appropriate intervals, and the number found in each interval is shown in the last column. The last column shows remarkably close agreement with the theoretical distribution given in the third column. Thus, by means of an exhaustive enumeration conducted on a small amount of data, a very close approximation has been obtained to the ideal distribution. Anyone who wishes to ascertain critical values for t by this cumbersome bit of arithmetic can do so, and he would have the assurance that the behavior of the t values was determined on the same experimental material and under the same circumstances under which he will later employ it as a guide. It is more convenient to use the mathematically determined critical values. The agreement of the observed values with the tabled ones would undoubtedly improve if this numerical exercise was tried on a larger scale.

The important thing is that it is possible, by performing a quite simple numerical operation on a set of measurements, to obtain a quantity t, about which predictions can be made using mathematical considerations alone. Critical values can be derived which, it can be predicted, will be exceeded any given small percentage of the time. The obtaining of a value of t in excess of this critical value leaves the experimenter with the option of believing either that a very improbable thing has happened, or that the hypothesis assumed (the null hypothesis) in getting the critical value is not true for his data. The rejection of the null hypothesis means accepting the alternative that a real effect has been demonstrated. No more penetrating approach exists. Infinite experience, which is what would be needed to get the exact mathematical limits, coupled with a mental capacity for appraising any particular collection of data in the light of this experience, can do no better than a novice with the simple formula for t and a statistical table.

The notion of making the division depend upon a *random* selection is fundamental to the application of the statistical method. The experimenter stands with the eight specimens before him on a table. Presumably he can tell nothing about the specimens. He might as well be blindfolded. Perhaps he should be at this stage. Once the specimens have been divided into two heaps the die is cast. It may be that the experimenter has obtained an unequal division which, depending upon which lot is assigned to the elevated temperature, may lead him to conclude either that an effect exists when in fact there is none, or that there is no effect because the effect has been counterbalanced by the unequal division. The point is that the risk of obtaining an unequal division can be calculated *provided that all thirty-five possible divisions were given an equal opportunity to occur.* This is commonly achieved by numbering the specimens, placing eight slips in a hat, and drawing out four of them. This is the blindfold the experimenter uses in order to play fair.

The discussion of the method of pairing off the specimens was left incomplete. Nothing was said about the way the two members of each pair were split between the two sets. If a coin is tossed to determine which member of the pair goes to the reference set, the random element is preserved. It is necessary only to modify the statistical procedure to make proper allowance for the pairing process. It has long been the practice of experimenters to employ such balancing schemes, knowing that this makes the sets more alike and hence more sensitive in the detection of an effect. Until 25 years ago no machinery existed for adapting the statistical treatment to take cognizance of the skill of the experimenter in his experimental procedures. The analysis of variance does provide a means of introducing into the interpretation of the data due recognition of the steps the experimenter has taken to achieve good experimental work.

In order to pass judgment upon experimental effects the arrangement of the experimental program must make provision for a valid estimate of the variance. This variance may be estimated from replicates or, in complex experiments, from multifactor interactions. In either event the experimental arrangement must include a randomizing operation, whether this takes the form of the assignment of specimens to test lots, the order in which the measurements are made, or some other respect appropriate to the study concerned. This randomization may be subject to restrictions which have been adopted for good experimental reasons. There have been developed a large number of ingenious types of restrictions which have as their purpose the improvement of the precision of experimental works. These are commonly discussed under the heading, "Design of Experiments." A number of simple types of design will be examined in the next chapter.

CHAPTER 9

Arrangements for Improving Precision

Grouping Measurements in Blocks

There are many opportunities in the laboratory to use the technique of paired observations. A sample may be weighed out and put into solution. Aliquots from this solution form natural pairs because they are obviously more alike than solutions prepared from different samples. As Student pointed out in 1926, the technique of pairing is not a modern invention but probably goes back to the Ark. It is easy to arrange matters so that more than two aliquots are available from the solution prepared from each sample. The number of aliquots taken will depend upon the objectives of the investigation. It may be that four modifications of an analytical procedure are under investigation. In such a case the data may be collected in a table similar to Table 27.

TABLE 27. TABULATION OF ANALYTICAL RESULTS

Sample No.	Analytical Procedure				Sample Ave.
	I	II	III	IV	
1	..	..	..	..	..
2	..	..	..	..	..
3	..	..	..	..	..
.					
.					
.					
n	..	..	..	..	..
Ave.	..	..	..	..	..

Examination of such a table may be made by directing attention to two of the columns and going back and forth between them down the list of samples, looking for any tendency of the values in one column to run higher than those in the other. It will be necessary to do this six times, as that is the number of pairings among four columns. If there are ten columns the number of pairings is 45, which means that $45n$ differences must be inspected and carried in the mind, as far as possible, in order to draw some conclusions.

It is much easier to examine the data using the technique of the analysis of variance (Table 28). The methods may be considered one

TABLE 28. ANALYSIS OF VARIANCE

Source	D.F.	S.S.	M.S.
Methods	3	..	a
Samples	$n - 1$	..	b
Interaction	$3n - 3$	..	c
Total	$4n - 1$	..	

factor, and the samples the other. The interaction between samples and methods provides the estimate of the variance of the measurements.

The use of the interaction is based upon the idea that the various biases of the four methods are independent of the sample examined, that the relationships among the methods persist over the whole array of samples. If there is any inclination to demur on this point, it is essential to stop and consider carefully just what are the objectives of the investigation. If one of the objectives is to ascertain whether or not there does exist an interaction between methods and samples, then some provision should be made for replicate measurements on the same sample by the same method. These replicates will yield an estimate of the variance of a measurement, and this in turn is used to judge the interaction variance. In the absence of interaction these two estimates have the same expectations. The existence of interaction leads to the statement that the difference between the methods depends upon the circumstances. It is a stronger statement to claim that one of the methods gives high results without adding the qualifying clause "depending upon the circumstances." Such a statement is often made if the method in question gives high results for every sample examined, even if the departure varies from sample to sample. Exceptions or borderline cases may be tolerated in view of the errors of measurement. This is equivalent to basing the statistical judgment of the difference between the averages upon the interaction variance. The F test is applied to the ratio a/c.

The objectives of the study determine the experimental program. The statistical evaluation must be fitted to the objectives. Notice, for example, that the claim of high results, irrespective of circumstances, carries with it the responsibility of securing an array of samples which may fairly be considered typical of those likely to be encountered in the future. Suppose that all the samples that came through the laboratory over a long period happen to have been saved, that a number of these are chosen at *random* from the shelves, and that these samples all show high results with the method in question. Most workers would say the method runs high without further qualification. The statistical test judges the *average* amount high by the constancy of this bias from sample to sample. It is clear enough that a very erratic amount of bias among the samples will lead to caution even if all samples happen to show a high result by the method. The caution displayed will depend upon the number of samples examined. The statistical table of F is a helpful guide in just such situations.

Nothing has been said about the assignment of the four aliquots from each sample to the four analytical methods. Chemists will be prone to consider the aliquots identical and to think that no consideration need be given this point. Often this may be true, and in any event there may have been introduced a certain amount of unintentional randomization. If four aliquots are drawn by a pipette, the beakers may be assigned to the methods without reference to the order in which they were taken. Very often, however, in a repeated operation, a systematic sequence of movements becomes established and the first aliquot drawn always goes to the same method. It is conceivable that the pipette adsorbs material when first filled, or that by the warmth of the hands the volume changes during the four withdrawals. If a systematic sequence is maintained, the particular effects that may be attached to the order of withdrawal are immediately transferred and credited to the methods. It is hardly worth arguing the point as to whether or not with due care such disturbances are avoided. It costs nothing to randomize the order, and this step effectively deals with the situation. Randomization does two things. It prevents some overlooked effect becoming identified with an experimental factor, and it insures that any small overlooked effects are impartially distributed among the comparisons used to judge the methods and the comparisons used to estimate the error variance. By imposing certain restrictions upon the randomization these suspected effects may be isolated and examined separately. If they are found to exist they give no cause for worry because the restricted randomization has removed their effect upon the comparisons of the methods.

In statistical parlance a complete set of all the different experimental trials is called a *replication*. One result by each of the four analytical methods just discussed forms a replication. It is the essence of good experimental work to arrange matters so that comparisons among the items comprising such a replication are made under the same conditions.

This is the reason that four aliquots from one sample are assigned to a replication. This is a natural extension of the established merit of using paired specimens. The term *randomized block* is commonly used to indicate a replication that has been assigned to a relatively homogeneous group of specimens. The formation of such groups may arise in the natural order of events, as in the case of the aliquots from one sample, or by the skillful inspection and grouping together of like items in the experimental material. Naturally familiarity with the material and circumstances is indispensable in recognizing the existence of such homogeneous groups.

The utility of randomized blocks goes far beyond the grouping of like specimens. All the above analyses by the four methods may be carried out on aliquots drawn from a large bottle of stock solution. In this case the classification by samples disappears entirely. It is still possible to arrange a grouping into blocks. The chemist will readily recognize that it is inadvisable to do all the analyses by one method on one day, by the next method on a later day, and so on. There are environmental factors that are often uncontrolled. The humidity, the daylight lighting, and the temperature are examples of conditions which may influence the results. Even air conditioning and artificial lighting do not provide "man-conditioning." The operator often varies from day to day in his judgment of end points or his method of reading the level of a meniscus on a burette. The events of a day, or a shorter period, form a natural group. This would lead to the performance of a complete replication as a randomized block, the time periods now taking the role formerly held by samples.

Certain it is that conclusions of any consequence about the relative merits of the methods must not depend upon what day the analyses were done. Any statements about the data should hold irrespective of the time the work was done. For this reason the interaction variance, days by methods, is the logical and reliable estimate of the variance of the measurements to be used in appraising the data. All the patient efforts to provide "controlled conditions" for the work are direct acknowledgments of this state of affairs. It is generally and erroneously assumed that the "controlled conditions" have accomplished the purpose intended: the reduction of the interaction variance to that obtained from duplicates run in parallel. This is seldom the case, and the precision of the comparison is usually improved by arranging the program of work in the form of randomized blocks.

Example 16. This example shows a comparison of five samples by a spectrographic method. The five samples were run on each of five plates so that there are five randomized blocks. The samples were taken in a random order for each plate. The order and the results appear in Table 29. The results are then tabulated in Table 30. The analysis of variance (Table 31) of these twenty-five measurements shows entries for samples, plates, and the interaction of samples with plates. It is immediately obvious that the variance associated with plates is much

TABLE 29. SPECTROGRAPHIC DETERMINATIONS OF NICKEL

Order	Plate I		Plate II		Plate III		Plate IV		Plate V	
1	B	2.02	A	1.80	A	2.08	A	1.94	C	2.01
2	A	2.05	C	1.79	B	2.19	E	1.99	D	2.02
3	E	2.11	B	1.79	E	2.11	B	1.98	B	2.02
4	D	2.04	D	1.77	C	2.08	C	2.03	A	2.00
5	C	2.06	E	1.86	D	2.15	D	2.03	E	2.02

TABLE 30. TABULATION OF DATA FOR COMPUTATION

Sample	Plate No.					Ave.
	I	II	III	IV	V	
A	2.05	1.80	2.08	1.94	2.00	1.974
B	2.02	1.79	2.19	1.98	2.02	2.000
C	2.06	1.79	2.08	2.03	2.01	1.994
D	2.04	1.77	2.15	2.03	2.02	2.002
E	2.11	1.86	2.11	1.99	2.02	2.018
Ave.	2.056	1.802	2.122	1.994	2.014	1.9976

larger than that associated with interaction, indicating that, in spite of every care to control conditions, runs made on the different plates are less comparable than runs made on the same plate. It turns out that the variance for samples is about the same as the variance for interaction, indicating that no differences have been established among the samples.

TABLE 31. ANALYSIS OF VARIANCE

Source	D.F.	S.S.	M.S.
Samples	4	0.005056	0.001264
Plates	4	0.287136	0.071784
Interaction	16	0.019264	0.001204
Total	24	0.311456	

Many experimenters find the interaction variance a rather mysterious method of estimating the precision of the measurements. If it is now disclosed that the five samples labelled A, B, C, D, and E are in fact all one sample, it is possible to estimate the precision by more familiar methods. The five measurements on each plate are from one sample and may be used to make an estimate of the variance based on four degrees of freedom. The estimate obtained from each plate is shown in Table 32.

TABLE 32. ESTIMATES OF VARIANCE

Plate	S.S.	D.F.	Variance
I	0.00452	4	0.00113
II	0.00468	4	0.00117
III	0.00908	4	0.00227
IV	0.00572	4	0.00143
V	0.00032	4	0.00008
Total	0.02432	20	0.001216

These estimates may be pooled by accumulating the sums of squares and dividing by the total degrees of freedom. The resulting variance is in good agreement with the variance computed for interaction. It may also be noted that in the original analysis of variance the entry for samples is in effect another estimate of the variance of the measurements, since the samples are known to be all alike. This estimate, when pooled with the estimate for interaction in Table 33, gives precisely the

TABLE 33. VARIANCE FOR SAMPLES AND INTERACTION

Source	S.S.	D.F.	Variance
Samples	0.005056	4	
Interaction	0.019264	16	
Total	0.024320	20	0.001216

result just obtained in Table 32 by accumulating the estimates of precision available from each plate. This is rather convincing evidence that the statistical technique of using the interaction really works. Would it work if the samples were different? This can be tested by adding to all the A values some constant, say 0.20, some other constant being added to all the B values, and so on. The reader is invited, nay urged, to perform this operation upon the data. If the analysis of variance is then made on the adjusted data, the estimate of the interaction variance will be found to come through unaltered from the value found on the original data. Evidently the interaction variance is not influenced by differences among the samples.

The formal examination of replicated blocks is identical with that followed for a two-factor experiment. Whereas it is often the case that two factors do interact with each other, it is not expected that an interaction will occur between factors and replicate blocks. The whole idea back of scientific measurements is that, if a certain relationship exists among the chemical methods (or materials) today, it can be observed again tomorrow. The work may involve the comparison of samples, one method being used for analysis. True, the relationships among the samples may alter with age, but if this is the case then time is a factor. Replicate blocks done at different times should be labelled with the dates of measurement. The numerical operations on the data are the same in either case; the interpretation changes.

The example of the five spectrographic plates shows that by insuring a complete replication on each plate the comparisons among the samples are as precisely achieved as if all twenty-five runs were crowded onto one plate. No difficulty arises because the five measurements of a sample are located in five different plates and no longer can be used directly to estimate the precision because of the differences among the plates. The variance due to the differences among the plates is segregated, using the plate averages for this purpose. These averages will represent the plates provided that the same five samples appear on each plate, that is, that each plate carries a replication.

The effectiveness of replicate blocks in improving the precision is demonstrated by imagining that the assignment of samples to plates had been entirely haphazard. The assignment of the samples might have been as shown in Table 34, certainly a very uneven distribution

TABLE 34. HAPHAZARD ASSIGNMENT OF SAMPLES TO PLATES

Plate I		Plate II		Plate III		Plate IV		Plate V	
A	2.02	E	1.80	B	2.08	C	1.94	A	2.01
A	2.05	D	1.79	B	2.19	D	1.99	C	2.02
E	2.11	D	1.79	E	2.11	A	1.98	D	2.02
B	2.04	C	1.77	B	2.08	D	2.03	C	2.00
B	2.06	A	1.86	E	2.15	E	2.03	C	2.02

Sample averages: A, 1.984; B, 2.090; C, 1.950; D, 1.924; E, 2.040.

among the plates. The same data are used because all the measurements are in fact on one sample, and identifying letters may be assigned to play the game of comparing the samples. In any real case it would not be known that the samples were alike (this is to be decided by examining the data); therefore the variance due to plates cannot be segregated as the plates do not have matching samples. No favoritism was shown in assigning the letters, and the question now resolves itself into deciding whether or not the sample averages differ more than would be expected, considering the variation of the replicate measurements on the samples. Statistically this is equivalent to comparing the variance between samples with the variance within samples. The comparison is shown in Table 35.

TABLE 35. ANALYSIS OF VARIANCE

Source	D.F.	S.S.	M.S.	F
Between samples	4	0.091016	0.022754	2.06
Within samples	20	0.220440	0.011022	
Total	24	0.311456		

As was to be expected in view of the known identity of the samples, the calculated F ratio is well within the predicted limit of variation-2.87, taken from the F table. This is a poorly conceived arrangement, and the consequences are reflected by the large variance, 0.011022, for the precision of the comparisons compared with the variance, 0.001204, found when replicated blocks were employed. The data can be validly interpreted because the assignment to plates was in a random manner and replicate runs of the *same* sample were given the same opportunity to vary (by being on the same or different plates) as were runs on *different* samples. Using the plates as blocks made it possible to pick up differences between samples one-third as large as is the case here. All in all this is not a happy arrangement, because the variance of the measurements is now a hodgepodge of the variance between duplicates on the same plate and the variance between plates. Unfortunately a lot of experimental work falls in this category.

The randomized block is the most simple of the arrangements available for improving the precision of comparisons. It is often extremely effective. The following instances are typical of the situations which call for the use of randomized blocks.

It is desired to determine the effect of r different tanning procedures on some of the properties of leather. The plan calls for taking r specimens from each of h hides. The r specimens from each hide are assigned at random, one to each tanning procedure. The number of hides, h, depends upon the number of replications deemed necessary. Comparisons between tanning methods now involve comparisons between specimens cut from the same hide, and these are known to be more alike than specimens from different hides. The precision will be much less if the specimens, $r \times h$ in number, are mixed up, losing the hide identity, before assigning them to the tanning process. The use of a hide as a source of homogeneous material to furnish specimens for the replication is bound to appeal to the laboratory man. It should be even more acceptable when it is realized that a statistical technique exactly appropriate for this scheme is available.

ARRANGEMENTS FOR IMPROVING PRECISION

Another example is afforded by the comparison, by means of a Geiger counter, of a number of radioactive materials. The counting rate is vulnerable to fluctuations in the voltage supply and probably in other conditions as well. The experimenter recognizes this, and a scheme often employed is to select one of the samples as a reference, R, and sandwich the other samples, A, B, and C, in between repeated determinations of the reference sample. The order in which the measurements are taken looks like this:

$$R\,A\,R\,B\,R\,C\,R\,A\,R\,B\,R\,C\,R\,A\,R\,B\,R\,C\,R, \text{ etc.}$$

Half the measurements are expended upon the reference standard by this program. The use of randomized blocks leads to the following order:

Block 1	Block 2	Block 3	Block 4	etc.
$C\,A\,B\,R$	$C\,B\,R\,A$	$A\,C\,R\,B$	$B\,A\,R\,C$	

The four measurements constituting a block are grouped together in respect to time, and fluctuations in voltage will be smaller than over the whole period for all the measurements.

Example 17. Howard H. Seliger reports an example of this technique ["A New Method of Radioactive Standard Calibration," *Journal of Research of The National Bureau of Standards*, **45**, 496 (1950)]. The net counts per second given in this paper have all been diminished by 25.00 in Table 36 to reduce the arithmetic. Each block is a compact

TABLE 36. CODED COUNTING RATES

Sample	Block 1	Block 2	Block 3	Block 4	Total
A	1.46	2.00	1.48	1.51	6.45
B	2.58	3.03	2.82	2.90	11.33
C	4.15	4.61	4.31	4.13	17.20
D	4.54	4.52	4.53	4.15	17.74
Total	12.73	14.16	13.14	12.69	52.72

time period. There is no necessity that the blocks follow each other immediately in time.

The usual computations appear in Tables 37 and 38. There are several interesting things to observe in these data. First, the F value for blocks exceeds the 5 per cent critical value of 3.86, indicating that there exist substantial differences among the counting rates in the different time blocks. The samples are, by the arrangement, compared within time blocks. Consequently the block contribution to the variation among the sixteen entries may be set aside. Second, the samples are obviously different as shown by an F value of about 300. Third, the average entry is approximately 3.0, to which must be added the 25.0 subtracted from the data, giving 28.0 as the average counting rate per second. The average number of counts observed in the counting period for each entry was about 25,600, so that the counting period was about 915 seconds. The error variance for 28.0 has been found to be 0.0238. If 28.0 is multiplied by 915 seconds, the approximate total count is obtained. Multiply the error variance by the square of this factor and obtain 20,000 for the variance of an entry. This variance

TABLE 37. CALCULATIONS FOR THE ANALYSIS OF VARIANCE

	16 Entries	Block Totals	Sample Totals	Grand Total
Sum of squares	195.6948	696.2542	780.5190	2779.3984
Divisor	1	4	4	16
Quotient	195.6948	174.06355	195.12975	173.7124
Subtract	173.7124	173.71240	173.71240	
Sum of squares	21.9824	0.35115	21.41735	

TABLE 38. ANALYSIS OF VARIANCE

Source	D.F.	S.S.	M.S.	F
Blocks	3	0.35115	0.1170	4.92
Samples	3	21.41735	7.1391	299
Error variance	9	0.21390	0.0238	
Total	15	21.98240		

is of the same order of magnitude as the count (25,600). Assuming that the radioactive disintegrations follow a Poisson law, the theoretical precision of such counts is known to be the square root of the number of counts. The square root of the count gives the standard deviation of the count. In this example the square root of the average count is 160, and the square root of the experimentally estimated variance is 144. This agreement is excellent; in fact, for this set of data the error estimate is somewhat better than theory. In the long run the measurements, even when perfectly controlled, cannot have a better precision than theory postulates. Fourth, the variance exhibited by counts when no provision is made to delete differences between blocks, that is, trends in the counting rate of the equipment, is revealed by a series of ten counts on the same radioactive source, also given in Seliger's paper. The ten counts, for equal time periods, are:

14,805 15,021 15,335 15,287 15,384
15,378 15,040 15,292 15,398 14,908

The average of these readings is 15,185, and the calculated variance is 48,402, or about 3 times the theoretical variance. The analysis of variance table for the sixteen counts also makes it possible to see the effect of dispensing with the block arrangement. After setting aside the sum of squares accounted for by samples there remains the comparison shown in Table 39. The variance for the total is twice the within

TABLE 39. COMPARISON OF BETWEEN AND WITHIN BLOCK VARIANCE

	D.F.	S.S.	M.S.
Between blocks	3	0.35115	
Within block error	9	0.21390	0.0238
Total	12	0.56505	0.0471

block variance. In this case drifts and other disturbances were not as pronounced as they were during the sequence of the ten measurements just examined.

Latin Square Arrangements

The next example leads into a situation which already may have come to the mind of the reader. The effect of aging time on the breaking strength of cement will be used as an example. Three different times are to be investigated. The experimenter will naturally look for sets of three specimens as homogeneous as possible. Suppose that three molds are available for forming the specimens. A batch of cement is mixed and poured into the molds, and the specimens removed as soon as they are set. The operation is repeated until three mixes have been prepared, making available nine specimens to be distributed among the three time periods. Which shall constitute the block—the three specimens from one mix, or the three specimens formed from the same mold? If the molds are believed essentially identical, the choice will naturally fall upon the mix, the specimens being distributed at random to the three times. If the dimensions are critical and there is confidence that the process of making a mix is highly reproducible, the choice will fall on the three specimens made by a given mold to constitute the block. Fortunately this choice does not have to be made.

Consider the nine specimens arranged in a square as in Table 40, showing the origin of each specimen. The specimens are identified by lower-case letters. If all the specimens were aged for the same time, a simple two-factor analysis of variance (Table 41) would reveal differences between the molds and between the mixes. The interaction of mixes and molds provides an estimate of the variance of the measurements. There is no reason to believe that molds and mixes do interact with each other. If there are differences among the molds, these should be exhibited equally well regardless of the mix used. Similarly, if a mix happens to give unusually good specimens, this should be exhibited in every column, that is, regardless of the mold. The relationships between the molds are independent of the mix and vice versa. In such a case the interaction is genuinely an estimate of the error variance of the measurements. It is extremely instructive to observe that by this statistical treatment an estimate has been obtained of the expected

TABLE 40. DESIGNATION OF CEMENT SPECIMENS

Mix	Mold No. 1	2	3
I	a	b	c
II	d	e	f
III	g	h	i

TABLE 41. ANALYSIS OF VARIANCE FOR CEMENT SPECIMENS

Source	D.F.	S.S.	Variance
Mixes	2	B	B/2
Molds	2	M	M/2
Molds × mixes	4	I	I/4
Total	8		

agreement of duplicate specimens formed from the *same* mix with the *same* mold. This is something impossible to get at by direct experiment. Furthermore, the comparison of the aging times can be made to have the same precision as if this were possible.

In the conduct of the actual experiment with the three times the assignment of specimens is guided by a restricted form of randomization. The three specimens from a mix are assigned at random to the three times, subject to the restriction that no mold and no mix is represented more than once at each aging time. Let the aging times be A, B, and C. An arrangement which satisfies the imposed restriction is indicated in Table 42. This is one of twelve possible arrangements which can be

TABLE 42. ASSIGNMENT OF TEST TIMES TO CEMENT SPECIMENS

Mix	1	2	3
I	a A	b B	c C
II	d B	e C	f A
III	g C	h A	i B

constructed. The random element is introduced by picking some one of these twelve possible ways of assigning the specimens. It is traditionally called a Latin square because Latin letters are usually used to identify the tests.

The sets of three specimens assigned to the aging times have now been balanced in respect to both mixes and molds. It is important now to observe that a constant amount can be added to each A, another constant to each B, etc., representing the effects of aging without disturbing the sum of squares associated with either mixes or molds. Only the interaction sum of squares is influenced by effects produced by the aging. The problem is to isolate from the interaction the effects introduced by aging. This is achieved by calculating the sum of squares associated with the totals for A, B, and C, the three aging times, and subtracting it from the sum of squares for interaction (Table 43).

TABLE 43. ANALYSIS OF VARIANCE

	D.F.	S.S.	Variance	F
Variance for time	2	T	$T/2$	T/I'
By difference	2	I'	$I'/2$	
Variance for interaction: molds × mixes	4	I		

The experimental arrangement has achieved an equality between sets assigned to aging times irrespective of possible mix or mold differences. These mix and mold differences do not enter into the experiment and need not appear in the statistical examination except in so far as they are used to get the sum of squares for interaction. They are not indispensable for this purpose, as other ways are available for calculating the interaction directly. It is, of course, often of interest to ascertain whether or not there did exist differences among molds or mixes, and these are judged by the residual interaction variance, $I'/2$. The experiment has, besides accomplishing the primary object of determining the effect of aging time, furnished valuable information regarding the interchangeability of molds and the reproducibility of the mixing process.

Example 18. Twenty-seven test specimens of cement were prepared from nine molds, using three mixes. The entries show the assignment of the specimens to the three aging times, A, B, and C, and the breaking load found for each specimen. The loads shown in Table 44 are in units of 10 pounds and have been diminished by 600 units.

TABLE 44. BREAKING LOADS FOR CEMENT SPECIMENS

	First Latin Square Mold No.			Second Latin Square Mold No.			Third Latin Square Mold No.		
Mix	1	2	3	4	5	6	7	8	9
I	A 60	B 379	C 722	B 438	C 713	A 74	C 729	A 48	B 451
II	B 470	C 767	A 61	C 774	A 24	B 466	A 52	B 453	C 724
III	C 720	A 74	B 430	A 28	B 391	C 722	B 410	C 676	A 34

The molds were segregated into groups of three, and each group used to form a Latin square arrangement of the specimens. Examine first of all the data in the first Latin square (Table 45).

TABLE 45. ANALYSIS OF VARIANCE FOR FIRST LATIN SQUARE

Source	D.F.	S.S.	M.S.
Mixes	2	3,135	1,568
Molds	2	258	129
Interaction: mix × molds; times	4	679,653	
Total	8	683,046	

The last four degrees of freedom in Table 45 are then partitioned, using the totals for the three times to obtain the sum of squares for time (Table 46).

TABLE 46. ANALYSIS OF VARIANCE

	D.F.	S.S.	M.S.
Times	2	677,350	338,675
Interaction	2	2,303	1,152
Total	4	679,653	

It is evident immediately, without computing the F ratios, that the mean squares for mixes and molds are not appreciably larger than the interaction mean square after the removal of the variance due to time. So far it seems that it was a needless precaution to balance the sets of specimens as to their origin by mix and mold. On the other hand, if the molds were less uniform or the preparation of the mixes not so well controlled, the precision of the comparison of times would still be as good as shown by the present interaction variance.

The numbers of degrees of freedom involved are all small. Since this was evident in advance of the work, the program was extended to make use of nine molds instead of the three required to form a Latin square. This made possible three Latin squares. These squares are not unrelated, however, because the same three mixes were used for all three squares. The collection of data may be conveniently examined as a whole in Table 47.

TABLE 47. ANALYSIS OF VARIANCE FOR ALL 27 SPECIMENS

Source	D.F.	S.S.	M.S.	F
Mixes	2	5,244.67	2,622.34	4.48
Molds	8	4,496.00	562.00	
Times	2	2,072,897.56	1,036,448.78	
Residue: mixes × molds	14	8,185.77	584.70	
Total	26	2,090,824.00		

The sums of squares are calculated from the three overall totals for the mixes, from the nine totals for the molds, and from the three overall totals for times. The sum of squares for residue is obtained by difference, after calculating the sum of squares for the twenty-seven measurements.

In the interpretation of this analysis of variance the large mean square for times merely reflects what is obvious from a glance at the data—that the strength of the cement is changing sharply during the early days after making the specimens. The larger-scale program has confirmed the conclusion drawn from the 3×3 Latin square that the molds are sensibly identical. The virtual equivalence of the mean square for molds and residual variance implies that any differences between the molds are inconsequential compared with other sources of variation that affect the strength of the specimens. The increased number of measurements has now revealed that the mixes are not as interchangeable as the molds. The F ratio, 4.48, exceeds the critical 5 per cent value and leads to the conclusion that the mixing procedure is not up to the standard of the molds in the matter of the reproducibility of the specimens. This is not a surprising state of affairs.

The residual variance mean square is 584.70, which gives the standard deviation of a single strength test as $\sqrt{584.70}$ or 24.2 10-pound units. The breaking load of a specimen has 95 per cent confidence limits of 242×2.145, or close to 500 pounds. This is applicable to comparisons made on the same mix. The average of three tests narrows these limits to $(242/\sqrt{3}) \times 2.145 = 300$ pounds.

The measurements in this work are good enough to detect differences in the breaking strength of specimens prepared from different mixes of the same cement. This may suggest to some that possibly the effect of aging may be different for the various mixes. In the analysis made of the data no consideration was given to the interaction of time and mix. There are four degrees of freedom associated with this interaction.

TABLE 48. ANALYSIS OF VARIANCE FOR CEMENT SPECIMENS

Source	D.F.	S.S.	M.S.
Mixes	2	5,244.67	2,622.34
Molds	6	2,873.11	478.85
Times	2	2,072,897.56	1,036,448.78
Time × mixes			
From molds	2	1,622.89	884.45
From residue	2	1,914.89	
Residue	12	6,270.88	522.57
Total		2,090,824.00	

As a consequence of the assignment of specimens only two degrees of freedom for this interaction are included in the fourteen degrees of freedom for residual variance. The other two are included in the eight degrees of freedom for molds. These degrees of freedom may be separated and are shown in Table 48.

The time × mix interaction mean square falls considerably short of significance, although it is suggestive that each of the two components separately is larger than the residual mean square. If there had been differences between molds, the two degrees of freedom from molds could not be used for the time × mix interaction.

The isolation of the above sums of squares for the time × mix interaction is left as an exercise for the reader. Success in this exercise may be taken as demonstrating a good comprehension of the analysis of variance technique. HINT: In the analysis of variance for the first Latin square the residual interaction sum of squares (2 D.F.) may be obtained directly by using the totals for the three subscripts shown in Table 49. Each subscript is balanced with respect to molds, mixes, and times.

TABLE 49. ASSIGNMENT OF SUBSCRIPTS

Mix	Mold		
	1	2	3
I	A_1	B_2	C_3
II	B_3	C_1	A_2
III	C_2	A_3	B_1

The assignment of specimens to the times was deliberately arranged to insure that, by the time three mixes had been used, every one of the molds had supplied a specimen for all three times. If there had turned out to be differences among the molds, these would not have interfered with the precision of the comparison of the breaking strength of cements aged for the three periods. This precaution proved to be needless, as the analysis shows no detectable differences among the molds. Now that the interchangeability of molds is known the results may be appraised from a simple point of view. The nine specimens from a mix were assigned in sets of three to the three times. The three mixes make available nine sets of triplicates for estimating the precision. Each set of triplicate measurements provides an estimate of the error variance with two degrees of freedom. The pooled estimate from all nine sets gives an error variance of 508.0, based upon eighteen degrees of freedom. This is very close to the estimate 522.6, based upon twelve degrees of freedom, when various segregations of the variance were made in order to examine the data for evidence of interactions. It is impressive to see how a complex schedule, consciously arranged to examine certain questions and to guard against loss of precision in case certain variables (mold dimensions and mixes) were not controlled, yields as satisfactory an estimate of the precision as is given by the direct examination of triplicate measurements.

In the illustration just discussed one experimental factor, time, has been studied, using an arrangement to attain statistically, in effect, a homogeneity of the experimental material which did not exist physically. It is not possible to get three specimens from the same mold using the same batch. Earlier examples have shown that the estimate of error variance obtained from interaction is the same as that obtained from duplicates if there is no interaction between the factors. In this example there is no ground for postulating the existence of an interaction between molds and batches.

It may be that the project concerns two experimental factors which are known not to interact with each other. In such a case the two factors can be allotted to the rows and columns, and some environmental condition can be represented by the Latin letters. A study of chemical cells used as a means of setting up a reference temperature offers an instance of this application. The sealed cells, containing about a pound of chemical, are heated just sufficiently to melt the chemical and then carefully insulated. The temperature of the triple point is maintained very closely for a day or so. The object is to compare a number of these cells. Considerable time is required for the resistance thermometers to attain thermal equilibrium when placed in wells extending into the chemical. Consequently only one thermometer can be employed on a given melting of the cell. Differences in readings found among a number of cells could also arise if the thermometers were not in agreement. When all the thermometers in turn are tried with one cell, the work spreads over several days, introducing possible variations in the bridge measurements that do not occur among readings made on the same day.

There is no reason whatever to postulate an interaction between cells and thermometers. If there is a ranking among the thermometers, this is certainly not an affair of the cells. If the cells set up widely different temperatures, the ranking of the thermometers may be in one order at a low temperature and in some other order at a high temperature. Inasmuch as the cells agree within 0.002 degree, this range of temperature can hardly be enough to reverse or even alter the relationship among the thermometers.

Example 19. The following arrangement was employed. The letters A, B, C, and D in Table 50 refer to four runs, each run being made on a separate day, using the pairing of thermometer and cell specified by the margins. The entries are the readings converted to degrees Centigrade. Only the third and fourth decimal places are given, as the readings agreed up to the last two places.

TABLE 50. PAIRING OF CELLS AND THERMOMETERS

Cell	I	II	III	IV	Totals	Ave.
1	A 36	B 38	C 36	D 30	140	35.0
2	C 17	D 18	A 26	B 17	78	19.5
3	B 30	C 39	D 41	A 34	144	36.0
4	D 30	A 45	B 38	C 33	146	36.5
Totals	113	140	141	114	508	
Ave.	28.25	35.0	35.25	28.5		

Examination of the analysis of variance (Table 51) for these data shows greater mean squares for cells, thermometers, and days than for residue. The largest mean square, that for cells, arises chiefly from cell 2, which was known to be filled with a less purified lot of the chemical.

TABLE 51. ANALYSIS OF VARIANCE

Source of Variance	D.F.	S.S.	M.S.
Between thermometers	3	182.5	60.8
Between cells	3	805.0	268.3
Between days	3	70.0	23.3
Residue	6	43.5	7.25
Total	15	1101.0	

The F ratio for thermometers, $60.8/7.25 = 8.39$, exceeds the tabulated 5 per cent critical value of 4.76 but is just short of the 1 per cent value of 9.78. The evidence is fairly strong that there are differences among the thermometers even after they have been calibrated and corrections determined for them. The F ratio for days does not attain the critical value, but it is suggestive that there is a day-to-day effect. The estimated standard deviation for a single measurement is $\sqrt{7.25}$ or 2.69. If the effect of days on the readings is not eliminated, the sum of squares for days must be pooled with the residue. The standard deviation would then be:

$$\sqrt{\frac{70.0 + 43.5}{6 + 3}} = \sqrt{12.6} = 3.55$$

ARRANGEMENTS FOR IMPROVING PRECISION

The Latin square arrangement did diminish the error of the comparison. The accumulation of data from other similar studies gave convincing evidence of a day-to-day effect on the measurements. Again it is physically not possible to make two satisfactory measurements, using two thermometers on the same cell on the same day, or to compare two cells with one thermometer. Such comparisons inevitably appeared to involve day-to-day disturbances as the measurements were made on separate days. The employment of this schedule permitted a marked improvement in the precision of the comparisons *without* any modification of apparatus or technique, or any additional measurements. The measurements were simply grouped following a thought-out scheme of pairing each thermometer with each cell.

A Latin square involves three factors, one for rows, one for columns, and one for the letters. The factors must be meaningful and of course laid out in advance. The square is not a device for sorting out data after the work has been done. Furthermore the row and column factors must not interact with each other.

It might occur to someone conducting an interlaboratory study to send four samples to each of four laboratories, A, B, C, and D, asking each one to run the samples by the methods indicated in the schedule given in Table 52.

TABLE 52. ASSIGNMENT OF SAMPLES AND METHODS TO LABORATORIES

Sample	Method			
	I	II	III	IV
1	A	B	C	D
2	D	C	B	A
3	B	D	A	C
4	C	A	D	B

Each laboratory now makes only four analyses, using one or another of the methods on the four samples. Difficulty will arise if there exists an interaction between samples and methods, or between laboratories and methods. The sixteen analyses scheduled are a sub-set of the sixty-four that would be required if every laboratory ran every sample by all methods. When all sixty-four measurements are available, it would be possible to examine the data for the presence of two-factor interactions, using the three-factor interaction with twenty-seven degrees of freedom as an estimate of the error variance. This supposes also that the various methods and laboratories have about the same precision, which may not be the case.

INCOMPLETE BLOCK ARRANGEMENTS 99

Latin squares of any size may be constructed. They impose a restriction in that all three factors must be employed at the same number of levels. The numbers of rows, columns, and letters must be equal, and this requirement does not always fit the needs of the experimenter. On the other hand a little adaptation of the experimental program, which usually contains a number of more or less arbitrarily settled choices, often makes it possible to consider whether or not there would be an advantage in employing the Latin square arrangement.

Incomplete Block Arrangements

In certain instances it is possible to get around the condition that there must be an equal number of classes in all three factors. It may be desired, for example, to intercompare seven thermometers. Suppose that it is possible to mount only three thermometers in the bath so that they can be read with an optical aid in an overall time period short enough to insure an absence of pronounced drift in the bath temperature. An arrangement is available which lends itself to this situation. The seven thermometers are represented in Table 53 by letters. There are

TABLE 53. SELECTION OF THERMOMETERS FOR RUNS

Order of Reading	Run No.						
	1	2	3	4	5	6	7
I	A	B	C	D	E	F	G
II	B	D	F	E	G	A	C
III	C	F	E	A	B	G	D

seven runs, each with three thermometers, making a total of twenty-one readings. The order of reading is taken into consideration by seeing to it that every one of the seven thermometers gets the first, the second, and the third readings. It would at first seem that trouble would be encountered if the bath settings are different for the seven runs. The selection of the seven triads has been guided with a view to overcoming this difficulty and, in fact, making it unnecessary to strive to reproduce exactly the temperature in each of the seven runs. It is not so easy to repeat the temperature, and any attempt to do so requires reading a reference thermometer which must remain in the bath for all runs.

Single out a particular thermometer, say C, and proceed to compare its readings with those obtained by the other thermometers. Thermometer C appear in runs 1, 3, and 7, and it will be observed that in these three runs all the other thermometers, A, B, F, E, D, and G, have also been read. It may happen that in run 1 the bath setting was high by an unknown amount, x, over the nominal test temperature, T. In that event the reading of thermometer A will be

$$T + x + a$$

where a is the unknown correction to be applied to thermometer A. Similarly, thermometers B and C in the same run give the readings $T + x + b$ and $T + x + c$. Doubling the reading for C and subtracting the sum of the readings for A and B yield

Run 1:

$$2(T + x + c) - (T + x + a) - (T + x + b) = 2c - (a + b) = c_1$$

The observed quantity c_1 is numerically equal to twice the correction for C minus the sum of the corrections for A and B. Obviously runs 3 and 7 provide observed quantities c_3 and c_7, relating the correction in C to the corrections in E and F, and D and G. Tabulate the results, including a run which requires the taking of no observations:

Run 1	$2c - a - b = c_1$
Run 3	$2c - e - f = c_3$
Run 7	$2c - d - g = c_7$
Run 0	$c - c \quad\; = 0$

$$7c - (a + b + c + d + e + f + g) = c_1 + c_3 + c_7$$

Divide by 7:

$$c - \text{average correction} = \tfrac{1}{7}(c_1 + c_3 + c_7) = \bar{c}$$
$$c = \bar{c} + \text{average correction for all thermometers}$$

By repeating these numerical operations seven calculated quantities, $\bar{a}, \bar{b}, \bar{c}, \bar{d}, \bar{e}, \bar{f}, \bar{g}$, are obtained, all of which, when added to the same, though unknown, average correction for all of them, will give the corrections a, b, c, d, e, f, and g for the individual thermometers. If the average correction is not known, the absolute corrections cannot be obtained, but the corrections will be known relative to one another, so that readings by one thermometer can be converted to equivalent readings on another thermometer.

It may be that a standard reference thermometer has been used to read the bath each time. In that event the sum of all twenty-one readings on the thermometers under test may be compared with 3 times the sum of the seven readings taken with the standard thermometer. The difference between these sums divided by 21 will give the average correction so that absolute corrections to the thermometers may be obtained.

Another, and perhaps more simple, way to get the absolute corrections is to replace one of the test thermometers by the standard thermometer. The absolute correction for the standard is known, and as all the corrections are known relative to each other, it is easy to obtain absolute corrections for all thermometers.

This type of arrangement makes it possible to intercompare a large number of items without the necessity of maintaining conditions constant longer than is required to compare a small number of items. The task of the experimenter is made easier, and the results are often better than they otherwise would be.

The analysis of variance technique also is capable of application to this type of arrangement. Certain modifications are necessary because a run no longer contains a complete replication. There must be no interaction between the three factors: here thermometers, runs, and order of reading. Arrangements that satisfy the required conditions are limited to certain numerical combinations, as here 7, 7, 3. Some other combinations are 7, 7, 4; 11, 11, 5; 13, 13, 4; 15, 15, 7; 16, 16, 6; 21, 21, 5. If there are twelve thermometers it is a simple matter to make use of the 13, 13, 4 scheme. One of the thermometers is assigned two identifying letters, and the readings solemnly carried out as if they were two thermometers. This provides two estimates of the correction for the same thermometer, and the concordance of these estimates is an additional assurance to anyone who wishes a visible confirmation that the scheme works.

These arrangements might be very useful in microanalytical work. Very often a standard compound is periodically analyzed as a check. There are cases in which it is desired to intercompare with the highest precision a group of compounds, possibly obtained in a fractionating process. Consecutive duplicate analyses show better agreement than duplicates separated in time. These arrangements make it possible to extend the work over several days, maintaining the precision of comparisons between analyses made in the same working period.

The original idea of these incomplete blocks is due to F. Yates [*Journal of Agricultural Science*, **26**, 434–455 (1936)]. It has been employed many times when the physical circumstances draw emphatic attention to the existence of groups of limited size which possess an unquestioned homogeneity among the items. Incomplete blocks have been used to compare diets when the number of diets is in excess of the number of animals available in single litters. The leaves from a single plant constitute a natural group. The number of leaves is limited and often less than some biological experiments require for a complete replication. A considerable number of these arrangements, commonly called balanced incomplete blocks, have been devised. A special group of them are known as Youden squares. The arrangements have been catalogued and discussed by Cochran and Cox in their book *Experimental Designs*, and by Fisher and Yates in their collection *Statistical Tables*.

Many other arrangements, some of quite complex character, are available. The purpose of this chapter has been to draw attention to some representative arrangements. The selection of arrangements which make the most of the experimental situation is a real test of the experimenter's skill in arranging the work.

Example 20. Testing laboratories have developed equipment which makes possible a simple technique for the intercomparison of thermometers. It is usual to arrange for a very slowly and steadily rising bath temperature. An array of thermometers is then read in sequence and as nearly as possible at equal intervals. The array is then read in reverse order. The averages of the two readings for each thermometer are taken as independent of the rise in bath temperature.

In laboratories not specially equipped to test thermometers it seems possible that the device of the small blocks would prove useful. The data for this example were obtained for illustrative purposes.

Seven thermometers, designated by the letters A, B, C, D, E, F, and G and set up in that order, were read as indicated below. The vertical lines show the subdivision into sets of three. The thermometers had scale divisions of one-tenth of a degree and were read to the third place with optical aid. The readings were made just above 30 degrees Centigrade, and for convenience only the last two places are entered. That is, the entry 56 means the thermometer reading was 30.056.

Order of reading the thermometers is as follows (parentheses indicate omitted readings):

$A\,B\,(C)\,D\,|\,E\,F\,(G)\,A\,|\,B\,C\,(D)\,E\,|\,F\,G\,(A)\,B\,|\,C\,D\,(E)\,F\,|\,G\,A\,(B)\,C\,|\,D\,E\,(F)\,G$

The sets of three are tabulated in Table 54 with the observed readings

TABLE 54. THERMOMETER READINGS

Order of Reading within Set	Set 1		Set 2		Set 3		Set 4		Set 5		Set 6		Set 7		Ave.
1	A	56	E	16	B	41	F	46	C	54	G	34	D	50	42.4
2	B	31	F	41	C	53	G	32	D	43	A	68	E	32	42.9
3	D	35	A	58	E	24	B	46	F	50	C	60	G	38	44.4
Uncorrected set ave.		41		38		39		41		49		54		40	
Corrected set ave.		34		36		43		45		45		48		51	

in thousandths of a degree. Each entry in the second row was read immediately after the entry above it. Consequently the average for the second row should be higher than the average for the top row by the rise in temperature during the time it takes to make *one* reading. Because of the omitted readings the entries in the last rows should be higher by the rise occurring in the time to take *two* readings. The averages for the three rows show rises of 0.0005 and 0.0015. Recognizing that the temperature rises only a very small amount in the time to take a reading, it is interesting to see that the averages are in the right order and fairly well spaced. The average rise in the interval of a reading is $(0.0444 - 0.0424)/3 = 0.0007$ degree. The sequence of readings has automatically compensated for the rise *within* a set because every thermometer is represented in the first, second, and third positions within the sets.

The temperature of the bath is, of course, rising steadily as the readings progress from set to set. The simple set averages will not show this very satisfactorily because the thermometers vary from set to set and presumably have different corrections. The measurements have, in fact, been made to determine these corrections. Before proceeding to determine the corrections it is worth commenting that the sequence in which the readings were taken leads to a remarkable symmetry. Not only is it possible to get the thermometer corrections, making allowance for the particular sets in which they were read, but it is possible to correct the set averages, making allowance for the fact that different triads of thermometers have been used in the sets. It is not necessary to know the thermometer corrections to achieve this.

Single out a particular set, say the first one, and proceed to compare it with the other sets. Through the medium of thermometer A set 1 may be compared with sets 2 and 6.

With thermometer A
 Twice set 1 $-$ (set 2 $+$ set 6) $= 0.112 - 0.126 = -0.014$
With thermometer B
 Twice set 1 $-$ (set 3 $+$ set 4) $= 0.062 - 0.087 = -0.025$
With thermometer D
 Twice set 1 $-$ (set 5 $+$ set 7) $= 0.070 - 0.093 = -0.023$

Notice that these results are independent of the thermometer corrections. All three results verify that the bath during set 1 was below the temperature of the other sets. Since set 1 was read first and the bath temperature was rising, this is to be expected.

By writing
$$\text{Set } 1 - \text{set } 1 = 0$$
and summing the four equations there is obtained

$$7 \text{ times temp. of set } 1 - \text{sum of temp. of all sets} = -0.062$$

or
$$\text{Temp. of set } 1 = \text{ave. temp. for all sets} - 0.009$$

The average of all twenty-one readings is 0.043 and is used as a datum.

Temp. of set 1 = 0.043 − 0.009 = 0.034

A similar set of operations gives the corrected temperatures for the other six sets. The corrected values show a very satisfactory upward trend without the reversals exhibited by the simple averages. Furthermore, the overall change from set 1 to set 7 is now found to be 0.017 degree. The set averages give the temperature at the middle of the set, that is, after two readings have been made. Consequently, during the course of 28 − 2 − 2, or 24, readings the bath has climbed 0.017 degree, or about 0.0007 degree per reading, which confirms the estimate previously found by comparing the averages for the order within sets. The rise in temperature from one set to the next is about 0.003, almost at the limit of reading. Nevertheless, the corrected set averages display this gradual rise in a most gratifying manner.

In the case of the sets the assumption of a fairly uniform rate of rise during the taking of the readings makes possible this check on the efficacy of a scheme that permits the comparison of the intervals even though they have been measured with different thermometers with unknown corrections. This should give confidence that the corrections to the thermometers may be obtained even if the intervals differ in temperature in a completely unknown way.

The computations may be made in a systematic manner as shown in Table 55.

TABLE 55. COMPUTATIONS FOR ADJUSTED VALUES

Ther.	1	2	3	4	5	6	7	3 Times Row Totals	Sum of Sets with Ther.	Diff.	$\frac{1}{7}$ Diff.	Adj. Value
A	56	58					68	546	399	147	21	64
B	31		41	46				354	364	−10	−1	42
C			53		54	60		501	427	74	11	54
D	35					43	50	384	389	−5	−1	42
E		16	24				32	216	353	−137	−20	23
F		41		46	50			411	386	25	4	47
G				32		34	38	312	406	−94	−13	30
Total	122	115	118	124	147	162	120	2724	2724	0		

Adjusted value for A

$$= \text{grand ave.} + \frac{3\Sigma \text{ readings with } A - \Sigma \text{ sets in which } A \text{ appears}}{7}$$

$$43 + \frac{546 - 399}{7} = 64$$

INCOMPLETE BLOCK ARRANGEMENTS

The final result is that there exists some temperature at which the readings of the thermometers would be:

$$A = 30.064 \qquad E = 30.023$$
$$B = 30.042 \qquad F = 30.047$$
$$C = 30.054 \qquad G = 30.030$$
$$D = 30.042$$

If the correction at 30 degrees is known for one of the thermometers, the corrections for the others are obtainable immediately.

The appropriate analysis of variance for these data is obtained as shown in Tables 56 and 57.

TABLE 56. COMPUTATIONS FOR ANALYSIS OF VARIANCE

	21 Readings	7 Set Totals	3 Order Totals	7 Diff.	Grand Total
Sum of squares	42,558	119,662	274,930	55,440	824,464
Divisor	1	3	7	21	21
Quotient	42,558.0	39,887.3	39,275.7	2,640	39,260.2
Subtract	39,260.2	39,260.2	39,260.2		
Sum of squares	3,297.8	627.1	15.5	2,640	

TABLE 57. ANALYSIS OF VARIANCE

Source	D.F.	S.S.	M.S.	S.D.
Sets	6	627.1	104.5	
Order	2	15.5	7.8	
Thermometers	6	2640.0	440.0	
Residue	6	15.2	2.5	1.6
Total	20	3297.8		

The chief interest here lies in the estimated standard deviation of the readings. The value 1.6, or 0.0016 degree, shows the skill of the operator. Single readings with a standard deviation of about 0.002 are impressive on thermometers graduated to tenths.

The formulas for the analysis of other incomplete block arrangements are available in the Introduction of Fisher and Yates's *Statistical Tables*.

CHAPTER 10

Experiments with Several Factors

Factorial Experiments

A research program that takes up the study of the effects produced by varying two or more factors provides the setting for what statisticians describe as a factorial experiment. Four factors, each at two levels, result in 2^4 or sixteen different experimental combinations of the factors. Increasing the number of factors and the number of levels runs up the total very fast. Until recently examination of the data by the analysis of variance technique has required that all the possible combinations be run. The large number of runs required frequently leads the experimenter to pick certain combinations, sometimes without much thought and sometimes because of an advance decision that certain combinations are not feasible in practice or will not give good results.

There is some difference in the points of view taken by the statistician and the experimenter. If the factors studied influence the yield of a chemical process, the experimenter searches through the results looking for combinations of the factors with high values. He is guided by individual results. The statistician looks at the data as a composite whole, seeking to establish direct effects produced by the factors and the

TABLE 58. ANALYSIS OF VARIANCE

Source	D.F.
Factor A	1
Factor B	1
Factor C	1
Factor D	1
Six two-factor interactions of the type $A \times B$	6
Four three-factor interactions, $A \times B \times C$	4
One four-factor interaction, $A \times B \times C \times D$	1
Total	15

influence that factors exert upon each other. The analysis of variance is the statistician's guide. Thus for four factors, each at two levels, the analysis of variance (Table 58) systematically explores the whole body of data.

The estimates of variance provide the key to the interpretation of the data. It is necessary to have data for all combinations to make this analysis of variance unless some rather special and tedious calculations are undertaken. In certain circumstances a systematic selection of the combinations lends itself to examination without excessive arithmetical work.

Confounding

From the viewpoint of the experimenter a valuable feature of multiple-factor programs is their frequent adaptability to subdivision into parts which offer an opportunity to improve the precision of the comparisons. The various combinations can be collected into groups, and these groups form natural experimental partitions of the whole program. It is necessary to hold experimental conditions constant only for the combinations grouped together. This is particularly important in a large-scale program. Often the whole program requires several batches of raw material, or extends over many days, or involves several pieces of equipment.

The smaller groups often can be performed using one batch of material, or on one day, or with just one piece of equipment. Combinations belonging to a group are in a favorable position in that comparisons between them will be more precise than if different lots of raw materials, or different machines, are involved. It is the peculiar merit of many factorial experiments that the whole program can be placed on essentially the same basis as if all the work were performed under the same constancy of conditions that apply within the groups. It is necessary to form the groups carefully so that certain requirements are met.

The rules for the formation of the groups or parts of the whole program are simple when the factors are all at two levels, or all at three levels. Other combinations are possible, and these are enumerated in a study by Yates (1937). The principles involved will be illustrated for factors at just two levels. The book by Cochran and Cox (1950) may also be consulted for examples showing factors at more than two levels.

It will be helpful to indicate the connection between the groups that are formed and the numerical operations that are involved in obtaining the sums of squares for the analysis of variance. The following scheme (Table 59) shows the sixteen possible combinations of four factors, A, B, C, and D, all of which can take on either of two levels indicated by the subscripts 0 and 1. In the sixteen cells there are entered identifying letter combinations. These range from (1), which stands for the low level of each factor, to *abcd*, which identifies the combination in which each factor is at the high level. It will be seen that the lower-case letter is used to show a factor at the high level, and the absence of the letter marks the factor at the low level.

TABLE 59. COMBINATIONS OF FOUR FACTORS

	B_0				B_1			
	C_0		C_1		C_0		C_1	
	D_0	D_1	D_0	D_1	D_0	D_1	D_0	D_1
A_0	(1) *	d	c	cd *	b	bd *	bc *	bcd
A_1	a	ad *	ac *	acd	ab *	abd	abc	abcd *

The analysis of variance may be computed in the usual manner by condensing the results to form subsidiary tables. It will be found that the sum of squares associated with factor A may also be obtained by taking the difference between the sum of the entries in the first row (A always at low level) and the sum of the entries in the bottom row (A at high level). All that is necessary is to square this difference and divide by 16. Similarly the sum of squares for the interaction A with B is based upon the difference between the sum of the first four items in the top row plus the last four items in the bottom row and the sum of the remaining eight entries. In turn every one of the main effects (A, B, C, D) and all the separate interactions depend upon a contrast of a group of eight properly chosen items with the remaining eight.

An asterisk has been inserted after eight of the combinations. These are, in this example, all the cases in which there is an even number of factors at the high level. The difference between the sum of the items marked with an asterisk and the sum of the remaining items, when squared and divided by 16, gives the sum of squares for the four-factor interaction $A \times B \times C \times D$. Interaction involving several factors usually gives estimates of the variance of the same order of magnitude as the error variance. The complex of mutual influences of several factors is of little scientific interest, and it would be a loss of no consequence if it were impossible to evaluate this interaction.

The scheme consists in dividing the whole program into two groups of eight as indicated by the asterisks. Each of these groups may then be assigned to a separate lot of raw material or performed on a different occasion. If this step introduces an effect upon the results, it will become identified with or *confounded* with the four-factor interaction. The sum of squares computed for this interaction may now become quite large, as it should if one of the lots of material used tends to give high results with all members of the group using this lot of material. This does not matter; it merely reflects the existence of a difference between the two lots of raw material. It is important to consider the impact of the introduction of two different batches of raw material on the other sums of squares associated with the direct effect of the factors and with the two- and three-factor interactions. If it be assumed that the difference between the two batches, as far as their influence on the yield is concerned, is the same regardless of the combinations of the experimental factors involved, then the effect of using two different batches is nil. Inspection of the A_0 and A_1 sets shows that the asterisks appear equally often in the two sets, so the effect of a difference between the batches cancels out. This is visibly true for the $A \times B$ interaction also, and closer study shows this balancing out of batch effects carries through for all the sums of squares associated with main effects, and two- and three-factor interactions. The only comparison affected is the four-factor interaction.

The whole program has been divided into two parts which can be performed independently. Effort can now be concentrated on maintaining comparable conditions for one half the entire program. If conditions change for the second half, no harm will be done unless these changes are themselves factors which interact with the experimental factors under study. Quite commonly the two parts are simply assigned to different times. The relationships among the experimental factors are presumably independent of the date of the experiment, yet circumstances may be connected with a given date which tend to influence in the same way all runs done on that date. The division of the program into parts avoids the necessity of trying to reproduce exactly all environmental circumstances on the two dates. Also it is easier to hold the environment constant for the time required to perform half the runs than for the whole program.

A study with three factors, each at two levels, may be subdivided into the two sets, *a, b, c, abc* and *ab, ac, bc*, (1). This confounds the three-factor interaction. The whole program will probably be performed in duplicate to give some idea of the reproducibility of the results. It is advisable to consider an alternative subdivision of the program for the duplicate series. The groups (1), c, ab, abc and a, ac, b, bc confound the two-factor interaction $A \times B$, leaving the contrast for $A \times B \times C$ available for evaluation. The program has now been subdivided into four parts. The analysis of variance is made in the usual way to get sums of squares for A, B, C, $A \times C$, and $B \times C$. The interaction $A \times B$ is evaluated using only the data in the first replicate, and $A \times B \times C$ using the data from the second set. Compact groups have been formed with a probable improvement in the comparisons at the expense of using only part of the data for two of the comparisons. The analysis of variance table will appear as in Table 60.

TABLE 60. ANALYSIS OF VARIANCE

Source	D.F.
Direct effect of A	1
Direct effect of B	1
Direct effect of C	1
Interaction $A \times C$	1
Interaction $B \times C$	1
Interaction $A \times B$ (1st rep.)	1
Interaction $A \times B \times C$ (2nd rep.)	1
Between parts of program	3
Duplicates	5
Total	15

One of the objections to a large-scale study is the extension of the time required for completion. The range of environmental conditions encountered over a long period is bound to be greater than that experienced in short intervals of time. Also it is more difficult to mix thoroughly large masses of reagents or raw materials than small batches. Everything points to the conclusion that greater precision will be obtained for comparisons among measurements taken in a compact, short-term investigation. This is exactly what has been achieved by the division of the whole program into parts by the device of confounding. The experimenter is able to select those comparisons that are of greatest interest or which it is vital to know precisely and make the assignment of the experimental runs favor these comparisons. Comparisons of little or no interest may be confounded completely, thus relinquishing all information about them, or partially confounded, as in the last illustration, so that less precise information is obtained for these comparisons. No changes in the operations are involved. There is required only a deliberate and considered programming of the order in which the experimental runs are made. The order should be randomized within each group.

This is the place to introduce an example of the gains that may result from performing the program in parts. Dramatic examples abound in the fields of biology, bioassay, medicine, and agriculture. Among chemists the reception so far accorded this experimental device is reminiscent of the dilemma of the young man seeking employment. No one will hire him until he has experience, and he cannot get this experience until someone gives him a job.

An experimental investigation of several factors which explores all combinations of these factors may, at first glance, seem out of the question because of the large number of runs involved. Against this deterrent there are three compensatory advantages. First, the analysis of variance will reveal which interactions are of no consequence. This permits combining the data to form subsidiary tables in which the entries are based on two or more measurements. The detection of effects by inspection is greatly helped by using these averages, which are more stable than the individual measurements. Second, there is no need to make duplicate runs because the multiple-factor interactions make available an adequate estimate of the error variance. Third, the partition of the program into carefully balanced parts diminishes this error variance because the comparisons which were the primary objectives of the investigation have been singled out for preferential consideration at the expense of comparisons of no particular interest.

A program including five factors at two levels for each factor will involve thirty-two runs. These thirty-two runs may conveniently be assembled into four groups, such as the following:

(1)	bc	abd	acd	abe	ace	de	$bcde$
ab	ac	d	bcd	e	bce	$abde$	$acde$
a	abc	bd	be	ce	ade	$abcde$	cd
b	c	ad	$abcd$	ae	$abce$	bde	cde

which confound the interactions $A \times B \times C$, $A \times D \times E$, and $B \times C \times D \times E$. All ten two-factor interactions are available for evaluation, as well as eight of the ten interactions with three factors. One of the five interactions involving four factors is confounded, but not the interaction involving all five factors. The analysis of variance is given in Table 61.

The experimental program has, in consequence, achieved an organization which permits a systematic and efficient examination of the data and has taken advantage of every opportunity to improve the quality of the work. The interpretation of the data rests upon the evidence obtained rather than upon the judgment of the worker going over the results piecemeal in an endeavor to see whether the data mean anything.

EXPERIMENTS WITH SEVERAL FACTORS

TABLE 61. ANALYSIS OF VARIANCE

Source	D.F.
Direct effects of factors	5
Interactions of two factors	10
Interactions of three factors (possibly also used for error)	8
Interactions of four factors } (error variance)	4
Interaction of five factors }	1
Between parts of the program	3
Total	31

Systematic Selection of Experimental Trials

A device has recently been proposed for curtailing the number of runs in large-scale factorial experiments. Suppose that there are six factors, each at two levels, making sixty-four runs. These can be split into two equal groups which represent the comparison associated with the six-factor interaction. The asterisks in the diagram mark the runs comprising one of the groups. The device is to perform only the runs marked

				A_0				A_1			
				B_0		B_1		B_0		B_1	
				C_0	C_1	C_0	C_1	C_0	C_1	C_0	C_1
D_0	E_0	F_0		*			*		*	*	
		F_1			*	*		*			*
	E_1	F_0			*	*		*			*
		F_1		*			*		*	*	
D_1	E_0	F_0			*	*		*			*
		F_1		*			*		*	*	
	E_1	F_0		*			*		*	*	
		F_1			*	*		*			*

with an asterisk, or alternatively the unmarked group, that is, to perform just half the program. Certain consequences flow from this planned amputation of the program. The six-factor interaction is gone forever. Each one of the six main effects is confounded with one or another of the six five-factor interactions, and each of the fifteen two-factor interactions is confounded with a four-factor interaction. Inasmuch as the four- and five-factor interactions tend to approximate the error variance, there is expected little, if any, inflation of the main effect and two-factor sums of squares. There are twenty three-factor interactions, and these are paired off in ten pairs, the members of each pair being confounded.

The analysis of variance in Table 62 shows the resulting situation.

TABLE 62. ANALYSIS OF VARIANCE

Source	D.F.
Direct effects of factors (6)	6
Two-factor interactions (15)	15
Three-factor interactions (10 pairs) used as estimate of the error variance	10
Total	31

An exploratory experiment of this type has been reported by Brownlee [*Annals of the N.Y. Academy of Sciences*, **52**: Art. 6, 820–826 (1950)] in a study of the factors influencing the production of penicillin.

The alternative to such a systematically contrived program is a hit-and-miss selection which inevitably will present tremendous problems as far as any statistical examination of the data is concerned. No one will quarrel with a point of view which has as its goal the holding of the amount of work to a minimum and the extraction of the maximum amount of information from the work done.

There is the possibility of drawing up schedules which allow for examination of the results at various stages in the work. Subsequent selection of the runs to be made will be guided by the outcome of the intermediate examinations. The idea may be illustrated using the same six-factor program just discussed.

The whole array of factors may be explored by performing the sixteen runs marked with the figure 1. Upon examination of the results of these runs the higher levels of A and D are considered eliminated as undesirable levels. A further four runs, those marked with a 2, together with four of the first runs, provide eight runs, which upon examination lead to the elimination of E at the higher level. As a final step, four more runs marked with the figure 3 provide a complete factorial for the remain-

			A_0				A_1			
			B_0		B_1		B_0		B_1	
			C_0	C_1	C_0	C_1	C_0	C_1	C_0	C_1
D_0	E_0	F_0	1	3	3	1				
		F_1	3	2	2	3	1			1
	E_1	F_0		1	1					
		F_1	2			2		1	1	
D_1	E_0	F_0					1			1
		F_1	1			1				
	E_1	F_0							1	1
		F_1		1	1					

ing factors, B, C, and F. The final choices between the levels for these three factors are made at this stage.

Further developments in the direction of using a fraction of a replication should be forthcoming if there is close coordination between statisticians and experimenters. Proposed programs can be passed upon by the experimenter with a view to the amount of work they involve and by the statistician with reference to their suitability for statistical examination. Actual examples of benefits obtained will be required to promote the widespread use of such systematic scheduling of experimental work. It has long been a guiding principle for the laboratory man to base the next step in an investigation upon the previously obtained results. Seldom have laboratory workers been willing to embark on an extensive factorial experiment which had to be completed before taking stock of the situation. The use of fractional replication may be the means of removing this objection to large-scale systematic programs.

List of Publications Referred To in the Text

Statistical Tables, Ronald A. Fisher and Frank Yates. Third Edition. Hafner Publishing Company, Inc., New York, 1948.

Statistical Methods for Research Workers, Ronald A. Fisher. Eleventh Edition. Oliver and Boyd, Edinburgh, 1950.

The Design of Experiments, Ronald A. Fisher. Fifth Edition. Oliver and Boyd, Edinburgh, 1949.

Experimental Designs, William G. Cochran and Gertrude M. Cox. John Wiley & Sons, Inc., New York, 1950.

Statistical Methods, George W. Snedecor. Fourth Edition. The Iowa State College Press, Ames, Iowa, 1948.

The Design and Analysis of Factorial Experiments, F. Yates. Technical Communication No. 35. Imperial Bureau of Soil Science, Harpenden, England, 1937.

APPENDIX

Table I   Critical Values of t
Table II   Critical Values of F at 5 Per Cent Level
Table III Critical Values of F at 1 Per Cent Level
Table IV Table of Squares

Tables I, II, and III are abridged from "Statistical Tables" by Ronald A. Fisher and Frank Yates. They are reproduced with the kind permission of the authors and their publishers, Oliver and Boyd, Ltd., of Edinburgh.

TABLE I. CRITICAL VALUES OF t

Per Cent Probability Level

D.F.	50	40	30	20	10	5	1
1	1.000	1.376	1.963	3.078	6.314	12.706	63.657
2	.816	1.061	1.386	1.886	2.920	4.303	9.925
3	.765	.978	1.250	1.638	2.353	3.182	5.841
4	.741	.941	1.190	1.533	2.132	2.776	4.604
5	.727	.920	1.156	1.476	2.015	2.571	4.032
6	.718	.906	1.134	1.440	1.943	2.447	3.707
7	.711	.896	1.119	1.415	1.895	2.365	3.499
8	.706	.889	1.108	1.397	1.860	2.306	3.355
9	.703	.883	1.100	1.383	1.833	2.262	3.250
10	.700	.879	1.093	1.372	1.812	2.228	3.169
11	.697	.876	1.088	1.363	1.796	2.201	3.106
12	.695	.873	1.083	1.356	1.782	2.179	3.055
13	.694	.870	1.079	1.350	1.771	2.160	3.012
14	.692	.868	1.076	1.345	1.761	2.145	2.977
15	.691	.866	1.074	1.341	1.753	2.131	2.947
16	.690	.865	1.071	1.337	1.746	2.120	2.921
17	.689	.863	1.069	1.333	1.740	2.110	2.898
18	.688	.862	1.067	1.330	1.734	2.101	2.878
19	.688	.861	1.066	1.328	1.729	2.093	2.861
20	.687	.860	1.064	1.325	1.725	2.086	2.845
21	.686	.859	1.063	1.323	1.721	2.080	2.831
22	.686	.858	1.061	1.321	1.717	2.074	2.819
23	.685	.858	1.060	1.319	1.714	2.069	2.807
24	.685	.857	1.059	1.318	1.711	2.064	2.797
25	.684	.856	1.058	1.316	1.708	2.060	2.787
26	.684	.856	1.058	1.315	1.706	2.056	2.779
27	.684	.855	1.057	1.314	1.703	2.052	2.771
28	.683	.855	1.056	1.313	1.701	2.048	2.763
29	.683	.854	1.055	1.311	1.699	2.045	2.756
30	.683	.854	1.055	1.310	1.697	2.042	2.750
40	.681	.851	1.050	1.303	1.684	2.021	2.704
50	.680	.849	1.048	1.299	1.676	2.008	2.678
60	.679	.848	1.046	1.296	1.671	2.000	2.660
120	.677	.845	1.041	1.289	1.658	1.980	2.617
∞	.674	.842	1.036	1.282	1.645	1.960	2.576

Table II. Critical Values of F at 5 Per Cent Level

D.F. Denominator	\multicolumn{10}{c}{D.F. Numerator}
	1	2	3	4	5	6	7	8	9	10	12
1	161	200	216	225	230	234	237	239	241	242	244
2	18.51	19.00	19.16	19.25	19.30	19.33	19.36	19.37	19.38	19.39	19.41
3	10.13	9.55	9.28	9.12	9.01	8.94	8.88	8.84	8.81	8.78	8.74
4	7.71	6.94	6.59	6.39	6.26	6.16	6.09	6.04	6.00	5.96	5.91
5	6.61	5.79	5.41	5.19	5.05	4.95	4.88	4.82	4.78	4.74	4.68
6	5.99	5.14	4.76	4.53	4.39	4.28	4.21	4.15	4.10	4.06	4.00
7	5.59	4.74	4.35	4.12	3.97	3.87	3.79	3.73	3.68	3.63	3.57
8	5.32	4.46	4.07	3.84	3.69	3.58	3.50	3.44	3.39	3.34	3.28
9	5.12	4.26	3.86	3.63	3.48	3.37	3.29	3.23	3.18	3.13	3.07
10	4.96	4.10	3.71	3.48	3.33	3.22	3.14	3.07	3.02	2.97	2.91
11	4.84	3.98	3.59	3.36	3.20	3.09	3.01	2.95	2.90	2.86	2.79
12	4.75	3.88	3.49	3.26	3.11	3.00	2.92	2.85	2.80	2.76	2.69
13	4.67	3.80	3.41	3.18	3.02	2.92	2.84	2.77	2.72	2.67	2.60
14	4.60	3.74	3.34	3.11	2.96	2.85	2.77	2.70	2.65	2.60	2.53
15	4.54	3.68	3.29	3.06	2.90	2.79	2.70	2.64	2.59	2.55	2.48
16	4.49	3.63	3.24	3.01	2.85	2.74	2.66	2.59	2.54	2.49	2.42
17	4.45	3.59	3.20	2.96	2.81	2.70	2.62	2.55	2.50	2.45	2.38
18	4.41	3.55	3.16	2.93	2.77	2.66	2.58	2.51	2.46	2.41	2.34
19	4.38	3.52	3.13	2.90	2.74	2.63	2.55	2.48	2.43	2.38	2.31
20	4.35	3.49	3.10	2.87	2.71	2.60	2.52	2.45	2.40	2.35	2.28
21	4.32	3.47	3.07	2.84	2.68	2.57	2.49	2.42	2.37	2.32	2.25
22	4.30	3.44	3.05	2.82	2.66	2.55	2.47	2.40	2.35	2.30	2.23
23	4.28	3.42	3.03	2.80	2.64	2.53	2.45	2.38	2.32	2.28	2.20
24	4.26	3.40	3.01	2.78	2.62	2.51	2.43	2.36	2.30	2.26	2.18
25	4.24	3.38	2.99	2.76	2.60	2.49	2.41	2.34	2.28	2.24	2.16
26	4.22	3.37	2.98	2.74	2.59	2.47	2.39	2.32	2.27	2.22	2.15
27	4.21	3.35	2.96	2.73	2.57	2.46	2.37	2.30	2.25	2.20	2.13
28	4.20	3.34	2.95	2.71	2.56	2.44	2.36	2.29	2.24	2.19	2.12
29	4.18	3.33	2.93	2.70	2.54	2.43	2.35	2.28	2.22	2.18	2.10
30	4.17	3.32	2.92	2.69	2.53	2.42	2.34	2.27	2.21	2.16	2.09
40	4.08	3.23	2.84	2.61	2.45	2.34	2.25	2.18	2.12	2.07	2.00
50	4.03	3.18	2.79	2.56	2.40	2.29	2.20	2.13	2.07	2.02	1.95
60	4.00	3.15	2.76	2.52	2.37	2.25	2.17	2.10	2.04	1.99	1.92
120	3.92	3.07	2.68	2.45	2.29	2.18	2.09	2.02	1.96	1.91	1.83
∞	3.84	2.99	2.60	2.37	2.21	2.09	2.01	1.94	1.88	1.83	1.75

Table III. Critical Values of F at 1 Per Cent Level

D.F. Denominator	D.F. Numerator										
	1	2	3	4	5	6	7	8	9	10	12
1	4052	4999	5403	5625	5764	5859	5928	5981	6022	6056	6106
2	98.49	99.00	99.17	99.25	99.30	99.33	99.34	99.36	99.38	99.40	99.42
3	34.12	30.81	29.46	28.71	28.24	27.91	26.67	27.49	27.34	27.23	27.05
4	21.20	18.00	16.69	15.98	15.52	15.21	14.98	14.80	14.66	14.54	14.37
5	16.26	13.27	12.06	11.39	10.97	10.67	10.45	10.29	10.15	10.05	9.89
6	13.74	10.92	9.78	9.15	8.75	8.47	8.26	8.10	7.98	7.87	7.72
7	12.25	9.55	8.45	7.85	7.46	7.19	7.00	6.84	6.71	6.62	6.47
8	11.26	8.65	7.59	7.01	6.63	6.37	6.19	6.03	5.91	5.82	5.67
9	10.56	8.02	6.99	6.42	6.06	5.80	5.62	5.47	5.35	5.26	5.11
10	10.04	7.56	6.55	5.99	5.64	5.39	5.21	5.06	4.95	4.85	4.71
11	9.65	7.20	6.22	5.67	5.32	5.07	4.88	4.74	4.63	4.54	4.40
12	9.33	6.93	5.95	5.41	5.06	4.82	4.65	4.50	4.39	4.30	4.16
13	9.07	6.70	5.74	5.20	4.86	4.62	4.44	4.30	4.19	4.10	3.96
14	8.86	6.51	5.56	5.03	4.69	4.46	4.28	4.14	4.03	3.94	3.80
15	8.68	6.36	5.42	4.89	4.56	4.32	4.14	4.00	3.89	3.80	3.67
16	8.53	6.23	5.29	4.77	4.44	4.20	4.03	3.89	3.78	3.69	3.55
17	8.40	6.11	5.18	4.67	4.34	4.10	3.93	3.79	3.68	3.59	3.45
18	8.28	6.01	5.09	4.58	4.25	4.01	3.85	3.71	3.60	3.51	3.37
19	8.18	5.93	5.01	4.50	4.17	3.94	3.77	3.63	3.52	3.43	3.30
20	8.10	5.85	4.94	4.43	4.10	3.87	3.71	3.56	3.45	3.37	3.23
21	8.02	5.78	4.87	4.37	4.04	3.81	3.65	3.51	3.40	3.31	3.17
22	7.94	5.72	4.82	4.31	3.99	3.76	3.59	3.45	3.35	3.26	3.12
23	7.88	5.66	4.76	4.26	3.94	3.71	3.54	3.41	3.30	3.21	3.07
24	7.82	5.61	4.72	4.22	3.90	3.67	3.50	3.36	3.25	3.17	3.03
25	7.77	5.57	4.68	4.18	3.86	3.63	3.46	3.32	3.21	3.13	2.99
26	7.72	5.53	4.64	4.14	3.82	3.59	3.42	3.29	3.17	3.09	2.96
27	7.68	5.49	4.60	4.11	3.78	3.56	3.39	3.26	3.14	3.06	2.93
28	7.64	5.45	4.57	4.07	3.75	3.53	3.36	3.23	3.11	3.03	2.90
29	7.60	5.42	4.54	4.04	3.73	3.50	3.33	3.20	3.08	3.00	2.87
30	7.56	5.39	4.51	4.02	3.70	3.47	3.30	3.17	3.06	2.98	2.84
40	7.31	5.18	4.31	3.83	3.51	3.29	3.12	2.99	2.88	2.80	2.66
50	7.17	5.06	4.20	3.72	3.41	3.18	3.02	2.88	2.78	2.70	2.56
60	7.08	4.98	4.13	3.65	3.34	3.12	2.95	2.82	2.72	2.63	2.50
120	6.85	4.79	3.95	3.48	3.17	2.96	2.79	2.66	2.56	2.47	2.34
∞	6.64	4.60	3.78	3.32	3.02	2.80	2.64	2.51	2.41	2.32	2.18

Table IV. Table of Squares

	0	1	2	3	4	5	6	7	8	9
0	0	1	4	9	16	25	36	49	64	81
1	100	121	144	169	196	225	256	289	324	361
2	400	441	484	529	576	625	676	729	784	841
3	900	961	1024	1089	1156	1225	1296	1369	1444	1521
4	1600	1681	1764	1849	1936	2025	2116	2209	2304	2401
5	2500	2601	2704	2809	2916	3025	3136	3249	3364	3481
6	3600	3721	3844	3969	4096	4225	4356	4489	4624	4761
7	4900	5041	5184	5329	5476	5625	5776	5929	6084	6241
8	6400	6561	6724	6889	7056	7225	7396	7569	7744	7921
9	8100	8281	8464	8649	8836	9025	9216	9409	9604	9801
10	10000	10201	10404	10609	10816	11025	11236	11449	11664	11881
11	12100	12321	12544	12769	12996	13225	13456	13689	13924	14161
12	14400	14641	14884	15129	15376	15625	15876	16129	16384	16641
13	16900	17161	17424	17689	17956	18225	18496	18769	19044	19321
14	19600	19881	20164	20449	20736	21025	21316	21609	21904	22201
15	22500	22801	23104	23409	23716	24025	24336	24649	24964	25281
16	25600	25921	26244	26569	26896	27225	27556	27889	28224	28561
17	28900	29241	29584	29929	30276	30625	30976	31329	31684	32041
18	32400	32761	33124	33489	33856	34225	34596	34969	35344	35721
19	36100	36481	36864	37249	37636	38025	38416	38809	39204	39601
20	40000	40401	40804	41209	41616	42025	42436	42849	43264	43681
21	44100	44521	44944	45369	45796	46225	46656	47089	47524	47961
22	48400	48841	49284	49729	50176	50625	51076	51529	51984	52441
23	52900	53361	53824	54289	54756	55225	55696	56169	56644	57121
24	57600	58081	58564	59049	59536	60025	60516	61009	61504	62001
25	62500	63001	63504	64009	64516	65025	65536	66049	66564	67081
26	67600	68121	68644	69169	69696	70225	70756	71289	71824	72361
27	72900	73441	73984	74529	75076	75625	76176	76729	77284	77841
28	78400	78961	79524	80089	80656	81225	81796	82369	82944	83521
29	84100	84681	85264	85849	86436	87052	87616	88209	88804	89401
30	90000	90601	91204	91809	92416	93025	93636	94249	94864	95481
31	96100	96721	97344	97969	98596	99225	99856	100489	101124	101761
32	102400	103041	103684	104329	104976	105625	106276	106929	107584	108241
33	108900	109561	110224	110889	111556	112225	112896	113569	114244	114921
34	115600	116281	116964	117649	118336	119025	119716	120409	121104	121801
35	122500	123201	123904	124609	125316	126025	126736	127449	128164	128881
36	129600	130321	131044	131769	132496	133225	133956	134689	135424	136161
37	136900	137641	138384	139129	139876	140625	141376	142129	142884	143641
38	144400	145161	145924	146689	147456	148225	148996	149769	150544	151321
39	152100	152881	153664	154449	155236	156025	156816	157609	158404	159201
40	160000	160801	161604	162409	163216	164025	164836	165649	166464	167281
41	168100	168921	169744	170569	171396	172225	173056	173889	174724	175561
42	176400	177241	178084	178929	179776	180625	181476	182329	183184	184041
43	184900	185761	186624	187489	188356	189225	190096	190969	191844	192721
44	193600	194481	195364	196249	197136	198025	198916	199809	200704	201601
45	202500	203401	204304	205209	206116	207025	207936	208849	209764	210681
46	211600	212521	213444	214369	215296	216225	217156	218089	219024	219961
47	220900	221841	222784	223729	224676	225625	226576	227529	228484	229441
48	230400	231361	232324	233289	234256	235225	236196	237169	238144	239121
49	240100	241081	242064	243049	244036	245025	246016	247009	248004	249001

Table IV. Table of Squares (Continued)

	0	1	2	3	4	5	6	7	8	9
50	250000	251001	252004	253009	254016	255025	256036	257049	258064	259081
51	260100	261121	262144	263169	264196	265225	266256	267289	268324	269361
52	270400	271441	272484	273529	274576	275625	276676	277729	278784	279841
53	280900	281961	283024	284089	285156	286225	287296	288369	289444	290521
54	291600	292681	293764	294849	295936	297025	298116	299209	300304	301401
55	302500	303601	304704	305809	306916	308025	309136	310249	311364	312481
56	313600	314721	315844	316969	318096	319225	320356	321489	322624	323761
57	324900	326041	327184	328329	329476	330625	331776	332929	334084	335241
58	336400	337561	338724	339889	341056	342225	343396	344569	345744	346921
59	348100	349281	350464	351649	352836	354025	355216	356409	357604	358801
60	360000	361201	362404	363609	364816	366025	367236	368449	369664	370881
61	372100	373321	374544	375769	376996	378225	379456	380689	381924	383161
62	384400	385641	386884	388129	389376	390625	391876	393129	394384	395641
63	396900	398161	399424	400689	401956	403225	404496	405769	407044	408321
64	409600	410881	412164	413449	414736	416025	417316	418609	419904	421201
65	422500	423801	425104	426409	427716	429025	430336	431649	432964	434281
66	435600	436921	438244	439569	440896	442225	443556	444889	446224	447561
67	448900	450241	451584	452929	454276	455625	456976	458329	459684	461041
68	462400	463761	465124	466489	467856	469225	470596	471969	473344	474721
69	476100	477481	478864	480249	481636	483025	484416	485809	487204	488601
70	490000	491401	492804	494209	495616	497025	498436	499849	501264	502681
71	504100	505521	506944	508369	509796	511225	512656	514089	515524	516961
72	518400	519841	521284	522729	524176	525625	527076	528529	529984	531441
73	532900	534361	535824	537289	538756	540225	541696	543169	544644	546121
74	547600	549081	550564	552049	553536	555025	556516	558009	559504	561001
75	562500	564001	565504	567009	568516	570025	571536	573049	574564	576081
76	577600	579121	580644	582169	583696	585225	586756	588289	589824	591361
77	592900	594441	595984	597529	599076	600625	602176	603729	605284	606841
78	608400	609961	611524	613089	614656	616225	617796	619369	620944	622521
79	624100	625681	627264	628849	630436	632025	633616	635209	636804	638401
80	640000	641601	643204	644809	646416	648025	649636	651249	652864	654481
81	656100	657721	659344	660969	662596	664225	665856	667489	669124	670761
82	672400	674041	675684	677329	678976	680625	682276	683929	685584	687241
83	688900	690561	692224	693889	695556	697225	698896	700569	702244	703921
84	705600	707281	708964	710649	712336	714025	715716	717409	719104	720801
85	722500	724201	725904	727609	729316	731025	732736	734449	736164	737881
86	739600	741321	743044	744769	746496	748225	749956	751689	753424	755161
87	756900	758641	760384	762129	763876	765625	767376	769129	770884	772641
88	774400	776161	777924	779689	781456	783225	784996	786769	788544	790321
89	792100	793881	795664	797449	799236	801025	802816	804609	806404	808201
90	810000	811801	813604	815409	817216	819025	820836	822649	824464	826281
91	828100	829921	831744	833569	835396	837225	839056	840889	842724	844561
92	846400	848241	850084	851929	853776	855625	857476	859329	861184	863041
93	864900	866761	868624	870489	872356	874225	876096	877969	879844	881721
94	883600	885481	887364	889249	891136	893025	894916	896809	898704	900601
95	902500	904401	906304	908209	910116	912025	913936	915849	917764	919681
96	921600	923521	925444	927369	929296	931225	933156	935089	937024	938961
97	940900	942841	944784	946729	948676	950625	952576	954529	956484	958441
98	960400	962361	964324	966289	968256	970225	972196	974169	976144	978121
99	980100	982081	984064	986049	988036	990025	992016	994009	996004	998001

INDEX

Accuracy, 6
Aliquots, 80, 82
Analysis of variance, 50
  precautions, 55
Analytical errors, 33
Ark, 80
Average, 8
Average deviation, 8
Averages, comparison of, 24, 31

Bales, sampling of, 37
Bartlett, M. S., 21
Bias, 7, 40, 41, 82
Biological experiments, 101
Blank, 40, 41
Blocks, grouping measurements in, 80
  randomized, 83
Brownlee, K. A., 113

Calcium determinations, 44
Carbon determinations, 22, 26
Cement tests, 64, 75, 90
Coal analyses, 35
Cochran, W. G., 73, 102, 107, 115
Colorimetric analysis, 6
Concrete, 72
Confidence limits, 18
  factors for, 18, 19
Confounding, 107
Cores, sampling, 37
Cox, Gertrude M., 73, 102, 107, 115
Critical values of t, 24, 25

Data, missing, 55
  requirements for, 72
  small sets, 3
Degrees of freedom, 13
Design of experiments, 79
Diets, comparison of, 101
Duplicate runs, 62
Duplicates, agreement between, 1
  omission of, 70

Eglof, W. K., 44
Ephedrine hydrochloride, 22, 26
Errors, analytical, 33, 54
  constant, detection of, 6, 40
  gross, 4
  sampling, 33
Expected value, 10
Experimental trials, selection of, 70

F ratio, 20
  upper limits, 21
F test, 29
  relation to t test, 50
Factorial experiments, 106
Factors for confidence limits, 18, 19
Fisher, R. A., 25, 50, 102, 105, 115, 117
Fit of straight line to data, 45

Geiger counter, 88
Gosset, W. S., 19

Hazel, W. M., 44
Hides, sampling of, 87

Incomplete block arrangements, 99
Interaction, 59, 64, 66, 70, 81, 108
  use as estimate of error, 62, 111
Intercept of straight line, 43
Interpretation of data, 1, 57, 61
Iron determination, 7

Latin square arrangements, 90, 96, 99
Lead determinations, 28
Least squares formulas, 41
Leather, experiments with, 74, 87
Leaves in biological experiments, 101
Line through the origin, 45
Litters as incomplete blocks, 101

m, mean, 11
Main effect, 66
Margin of error, 68

Material, homogeneous, 87
Mean, 10
  square, 46
Measurements, 1
  grouping in blocks, 80
  independence of, 6
Median, 9
Microanalytical determinations, 22
Missing data, 55
Mix, cement, 64, 90
Model, statistical, 18
Molds for specimens, 90
Mu, 10

Nickel determinations, 17
Normal distribution, 5, 11
Normal law of error, 11
Null hypothesis, 25, 78

Observations, paired, 28, 80

Paired values, 28, 80
Pairs, 9, 28
Parameter, 11
Photographic plates, 83
Planning experimental programs, 72
Poisson law, 89
Pooling information, 12, 21
Power, F. W., 22
Precision, 6
  combination of estimates, 10
  comparison of estimates, 10
  measure of, 8, 10
  of averages, 17
  of single measurements, 17
Prediction, 1, 2, 3
  erroneous, 19

Radioactive materials, 88
Random arrangements, 75
Randomization, 82
Randomized block, 83
Range, 24
Recrystallization, 66
Replication, 82

Replication of measurements, 72
Risk of errors in judgment, 20
Rounding off data, 7, 16

s, 11
Samples, small, 19
Sampling, errors, 33
  in two stages, 37
  schedule, 34
Sand lot selection of teams, 75
Scales, 73
Seliger, H. H., 88, 89
Sigma, 10
Slope of straight line, 41
Slopes, comparison of, 47
Snedecor, George W., 115
Specimens, allocation of, 74
Spectrographic determinations, 15, 83
Standard deviation, 10, 20
  calculation, 12
  of intercept, 43
  of slope, 42
Straight line, 40
Student, 19, 25, 27, 80
Sum of squares, 13
  calculation, 16

t test, 24, 44, 50
Tanning of leather, 87
Temperature, reference, 96
Thermometers, 96
  comparison of, 99, 102
Three factor experiments, 68
Tomkins, S. S., 35

Variance, 10
  test for homogeneity, 21, 23
  use to compare averages, 53

Wool sampling, 37, 38

Yates, F., 101, 102, 105, 107, 115, 117
Youden squares, 102

Zinc determinations, 15